CATALYSIS BY METALS

Les Houches School, March 19-29, 1996

Editors

Albert Jean RENOUPREZ
Hervé JOBIC

Springer-Verlag Berlin Heidelberg GmbH

Centre de Physique des Houches

Books already published in this series

1   Porous Silicon Science and Technology
         Jean-Claude VIAL and Jacques DERRIEN, Eds. 1995

2   Nonlinear Excitations in Biomolecules
         Michel PEYRARD, Ed. 1995

3   Beyond Quasicrystals
         Françoise AXEL and Denis GRATIAS, Eds. 1995

4   Quantum Mechanical Simulation Methods for Studying Biological Systems
         Dominique BICOUT and Martin FIELD, Eds. 1996

5   New Tools in Turbulence Modelling
         Olivier MÉTAIS and Joel FERZIGER, Eds. 1997

Book series coordinated by Michèle LEDUC

Editors of "Catalysis by metals" (N° 6)

Albert Jean Renouprez and Hervé Jobic

Institut de Recherches sur la Catalyse, UPR 5401 du CNRS, 2 avenue Albert Einstein, 69626 Villeurbanne cedex, France

The organisation and financial support of this school was provided by the Formation Permanente of the CNRS, under the supervision of Jean-Louis Portefaix.

ISBN 978-3-540-63708-0   ISBN 978-3-662-06221-0 (eBook)
DOI 10.1007/978-3-662-06221-0

© Springer-Verlag, Berlin, Heidelberg 1997
Originally published by Springer-Verlag Berlin Heidelberg New York in 1997

LECTURERS

Barbier J., Laboratoire de Catalyse en Chimie Organique, URA 350 du CNRS, Université de Poitiers, 40 avenue du Recteur Pineau, 86022 Poitiers, France

Bergeret G., Institut de Recherches sur la Catalyse, CNRS, 2 avenue Albert Einstein, 69626 Villeurbanne cedex, France

Binet C., Laboratoire Catalyse et Spectrochimie, UMR 6506, ISMRA, Université de Caen, 6 boulevard du Maréchal Juin, 14050 Caen, France

Chollier M.J., Laboratoire de Catalyse en Chimie Organique, URA 350 du CNRS, Université de Poitiers, 40 avenue du Recteur Pineau, 86022 Poitiers, France

Epron F., Laboratoire de Catalyse en Chimie Organique, URA 350 du CNRS, Université de Poitiers, 40 avenue du Recteur Pineau, 86022 Poitiers, France

Forissier M., Laboratoire de Génie des Procédés Catalytiques, CPE Lyon, 43 boulevard du 11 Novembre 1918, B.P. 2077, 69616 Villeurbanne, France

Gallezot P., Institut de Recherches sur la Catalyse, CNRS, 69626 Villeurbanne cedex, France

Jobic H., Institut de Recherches sur la Catalyse, CNRS, 2 avenue Albert Einstein, 69626 Villeurbanne cedex, France

Laffite E., LACCO, Université de Recherche Associée au CNRS, DO 350, Université de Poitiers, 40 avenue du Recteur Pineau, 86022 Poitiers cedex, France

Lavalley J.C., Laboratoire Catalyse et Spectrochimie, UMR 6506, ISMRA, Université de Caen, 6 boulevard du Maréchal Juin, 14050 Caen, France

Maugé F., Laboratoire Catalyse et Spectrochimie, UMR 6506, ISMRA, Université de Caen, 6 boulevard du Maréchal Juin, 14050 Caen, France

Micheaud C., LACCO, Université de Recherche Associée au CNRS, DO 350, Université de Poitiers, 40 avenue du Recteur Pineau, 86022 Poitiers cedex, France

Rohart E., LACCO, Université de Recherche Associée au CNRS, DO 350, Université de Poitiers, 40 avenue du Recteur Pineau, 86022 Poitiers cedex, France

Simon D., Laboratoire de Chimie Théorique, École Normale Supérieure de Lyon, 46 allée d'Italie, 69364 Lyon cedex 07, France, and, Institut de Recherches sur la Catalyse, CNRS, 2 avenue Albert Einstein, 69626 Villeurbanne cedex, France

Tourillon G., Laboratoire de Cristallographie-CNRS, 25 boulevard des Martyrs, B.P. 166, 3800 Grenoble cedex 09, France

Travers Ch., Institut Français du Pétrole, B.P. 311, 92506 Rueil-Malmaison cedex, France

Tréglia G., C.R.M.C.[2]-C.N.R.S., Campus de Luminy, Case 913, 13288 Marseille cedex 9, France

PREFACE

Catalytic reactions on metals are still nowadays involved in more than half of the chemical industrial processes. The winter school held at "l'École de Physique des Houches" in March 1996, 13 years after the first one, accounts for an evolution of the field in several directions.

First, the emulation between theoretical chemistry and solid state physics has emerged on heuristic concepts, leading not only to explanations of the observed phenomena but, for the first time, to predictions of the reactivity of catalytic systems and of the reaction pathways.

The second domain which during these years has become of primary importance is the abatement of the pollution. It concerns not only the conversion of polluting effluents but more and more major modifications of the processes to avoid the production of undesired products. Two striking examples are the necessary catalytic conversion of the 100 000 cubic meter of hydrogen that would be produced in a major incident of a nuclear power plant and the replacement of the CFC.

The valorization of agricultural supplies can already be considered as one of the major achievement of catalysis. Indeed, the carbon of biosustainable raw materials represents more than 2 orders of magnitude the amount extracted from fossil fuels each year. Moreover, the molecules are already highly functionalised in contrast with hydrocarbons which require costly steps to be converted to the same products. They are now of current use in the elaboration of cosmetics, vitamins, polymers, etc.

In spite of the large artillery of physical methods now available to characterize the catalysts, their preparation is still a bolt and most scientists consider it as purely empirical. The reason is probably because very few can be employed *in situ* or "on line" during the various preparation steps. Indeed, a powerful synchrotron or a High Resolution Electron Microscope are not available in everyday's laboratory experiment, but electrochemistry has recently proved to be able to monitor this elaboration.

Finally the main contest of the future years is certainly the obtainment of these famous "Taylor made catalysts" able to deliver at one's will any chiral compound. Obviously a large part of the third school of **catalysis by metals** in ... 2006 will be devoted to supramolecular catalysis.

A. J. Renouprez and H. Jobic

CONTENTS

LECTURE 1

IR Characterization of Metal Catalysts
Using CO as Probe Molecule

by F. Maugé, C. Binet and J.C. Lavalley

1. Introduction	1
2. Molecules adsorbed on a metal surface	1
2.1. Electric field	1
2.2. Selection rule: an application	3
2.3. IR characterization of adsorbed molecules	4
3. CO Adsorption	5
3.1. Electronic interaction between the CO molecule and transition metals	5
3.2. Frequency shifts for varying surface coverage	7
3.2.1. Vibrational coupling (or dynamic effect)	7
3.2.2. Chemical shift (or static effect)	8
3.2.3. Separation between dynamic and static effects	8
3.2.4. Special case for column IB metals	9
3.3. Structure of the adsorbed molecule layer	9
3.4. Influence of surface defects	12
3.5. Bimetallic catalysts	14
4. Examples of characterization of supported Pd Catalyst	14
4.1. Evaluation of the relative amount of crystallographic faces on Pd/Al$_2$O$_3$	14
4.2. Formation of carbonyl clusters by CO adsorption on highly dispersed Pd/CeO$_2$	16

LECTURE 2

Prolegomena to Magnetic Circular Dichroism in X-Ray Absorption Spectroscopy

by Ch. Travers

1. Introduction 19
2. Catalysts preparation.. 19
   2.1. Generalities... 19
   2.2. Theoretical notions 20
      2.2.1. Young-Laplace law 21
      2.2.2. Kelvin and Ostwald laws... 22
      2.2.3. Supersaturation. Nucleation and growing of particles.......... 23
         2.2.3.1. Influence of the supersaturation on nucleation and particles growing 24
   2.3. Review of the main elementary steps 27
      2.3.1. Thermal or hydrothermal treatments............................ 27
      2.3.2. Forming 29
         2.3.2.1. Forming microgranules 30
         2.3.2.2. Forming granules... 30
      2.3.3. Impregnation of active agents on a support.................. 33
         2.3.3.1. Impregnation without interaction 33
         2.3.3.2. Impregnation with interactions by ionic exchange. 36
      2.3.4. Unit operations after impregnation.............................. 40
3. Conclusion: example of preparation of industrial catalysts 42

LECTURE 3

Measurement of Catalyst Performances at the Laboratory

by M. Forissier

1. Catalyst testing and kinetic study 45
   1.1. Catalyst testing... 45
   1.2. Kinetic study... 46
2. Some definitions 46
   2.1. Complex chemical reactions.. 46
   2.2. Elementary step 46

2.3. Kinetic expression 47

2.4. Reaction rates.. 47

2.5. Conversion, yield and selectivity...................................... 48

2.6. Time on stream and residence time 48

3. Chemical or diffusion limitations 48

3.1. In a differential flow reactor 49

3.2. In an integral flow reactor... 49

4. Experimental devices for kinetic studies 50

4.1. Global point of view 50

4.1.1. Choice of the experimental conditions of the study 50

4.1.2. The control and the knowledge of the reactions conditions .. 50

4.1.3. Some experimental difficulties to detect 50

4.1.4. How to obtain practically the chosen reaction conditions 50

4.2. Some laboratory reactors often used................................. 50

4.2.1. The batch reactor 50

4.2.2. Fixed bed flow reactor without recirculation 52

4.2.3. Particular case: differential flow reactor 53

4.2.4. Perfectly mixed flow reactor 53

4.2.5. Pulse reactor 55

4.2.6. TAP (Temporal Analysis of Products)............................... 55

4.2.7. Laboratory reactors for triphasic reactions
with gas-, liquids- and solid catalyst.................................. 56

5. Reaction scheme 56

6. Kinetic model.. 57

7. Catalyst deactivation... 57

7.1. Deactivation causes... 58

7.2. Model of the deactivation 58

7.2.1. Hypothesis of the independence of the deactivation
and of the reaction 58

7.2.2. Fouling kinetic.. 59

7.3. Do not miss the deactivation.. 61

8. Determination of the kinetic constants 63

8.1. Linearisation method 63

8.2. Non-linear adjustment techniques 63

9. Interpretations of kinetic constant values 63

10. Detailed mechanism and site simulation................................. 65

11. Conclusions 65

LECTURE 4

Electronic Structure of Metals and Alloys: from Bulk to Surfaces and Clusters

by G. Tréglia

1. Introduction 67
2. Bulk electronic structure 68
   2.1. Pure metals 68
      2.1.1. One electron approximation (ab initio methods) 68
         2.1.1.1. Hartree(-Fock) approximation 68
         2.1.1.2. Local density functional approximation (LDA) 70
      2.1.2. Normal metals (jellium, pseudopotentials) 71
      2.1.3. Transition metals (tight-binding approximation) 72
      2.1.4. Energetics and structure 76
         2.1.4.1. Normal metals 76
         2.1.4.2. Transition metals 78
   2.2. Alloy $A_c B_{1-c}$ 81
      2.2.1. Electronic structure 81
      2.2.2. Mixing energy 83
3. Electronic structure of surfaces 85
   3.1. Pure metals 85
      3.1.1. Electronic structure 85
         3.1.1.1. Normal metals: surface states 85
         3.1.1.2. Transition metals 85
      3.1.2. Surface energy 89
         3.1.2.1. Normal metals 89
         3.1.2.2. Transition metals 89
      3.1.3. Atomic structure (transition metals) 90
         3.1.3.1. Surface relaxation 90
         3.1.3.2. Surface reconstruction 91
   3.2. Bimetallic systems 92
      3.2.1. Alloy surfaces: $A_c B_{1-c}$ 92
      3.2.2. Surface alloy: A/B 95
      3.2.3. Atomic superstructure 95
      3.2.4. Local order and densities of states 96
4. Electronic structure of clusters 96

4.1. Pure metals.. 96

    4.1.1. Electronic structure 96

    4.1.2. Atomic structure.. 97

4.2. Bimetallic systems 98

    4.2.1. Finite size effect on surface segregation 98

    4.2.2. Local densities of states and preferential sites...................... 100

LECTURE 5

Étude des adsorbats moléculaires sur surface métallique par spectroscopies de photoémission et d'absorption X

by G. Tourillon

1. Spectroscopie de photoémission XPS et UPS... 104

2. Spectroscopie d'absorption X-NEXAFS 105

3. Spectroscopie de déexcitation (Auger résonant et Raman résonant)....... 111

LECTURE 6

Catalysis by Metals: Contribution of Electrochemistry

by J. Barbier, M.J. Chollier and F. Epron

1. Preparation of bimetallic catalysts
   by surface electrochemical reactions 113

    1.1. Direct redox reactions in the preparation of bimetallic catalysts...... 113

    1.2. Redox reactions of adsorbed species in the preparation
        of bimetallic catalysts (« Recharge method »).............................. 115

    1.3. Catalytic reduction in the preparation of bimetallic catalysts........... 115

    1.4. Underpotential deposition... 116

    1.5. Concluding remarks.. 117

2. Characterization of metallic catalysts by the means
   of electrochemical methods 117

    2.1. Determination of metallic surface area by cyclic voltammetry 119

2.2. Thermodynamic study of the adsorption of different compounds by cyclic voltammetry 119

    2.2.1. Thermodynamic study of hydrogen adsorption on platinum 119

    2.2.2. Thermodynamic study of sulphur adsorption on platinum.... 122

    2.2.3. Thermodynamic study of maleic acid adsorption 124

2.3. Electrochemical potential of the catalyst during hydrogenation reactions in liquid phase 126

2.4. Concluding remarks 129

LECTURE 7

Catalysis and Automotive Pollution Control

by J. Barbier, C. Micheaud, E. Rohart and E. Lafitte

1. Introduction 133

2. General aspects 134

3. Catalytic combustion 135

  3.1. Catalytic oxidation of carbon monoxide 135

    3.1.1. Adsorption of carbon monoxide 135

    3.1.2. Adsorption of oxygen 137

    3.1.3. Experimental approach of catalytic oxidation of CO 138

  3.2. Catalytic oxidation of hydrocarbons (HC) 139

    3.2.1. Alkane oxidation 139

    3.2.2. Alkene oxidation 140

4. Role of water in the CO and unburnt hydrocarbons combustion 141

  4.1. Water gas shift reaction 142

  4.2. Hydrocarbon steamreforming reaction 143

5. NO_x reduction 143

  5.1. Conversion of NO_x in Three-Way Catalysts (TWC) 143

    5.1.1. Reduction of NO by CO 143

    5.1.2. Reduction of NO by H_2 145

    5.1.3. Reduction of NO by unburnt hydrocarbons 145

  5.2. Reduction of NO_x under lean operating conditions - Application to the diesel engine 146

    5.2.1. General aspects 146

    5.2.2. Fundamental approach 148

6. Concluding remarks 150

LECTURE 8

Chemisorption Bonds at Transition Metal Surfaces: Orbital Approach

by D. Simon

1. Introduction 153
2. The concepts 153
   2.1. Definition of local density of states 153
   2.2. Electron population and transfer 156
   2.3. Mean energy 156
   2.4. Overlap population 156
   2.5. Atomic properties 157
3. Illustration of the concepts 157
4. Interaction diagrams 161
5. Conclusion 164

LECTURE 9

Characterization of Metallic Catalysts by X-Ray and Electron Microscopy Techniques

by G. Bergeret

1. Definitions and generalities 167
   1.1. Particle size: mean diameters 167
   1.2. Model particles: total number of atoms and size 168
   1.3. Dispersion. Surface area 169
   1.4. Coordination number 169
2. X-ray techniques 170
   2.1. X-Ray powder Diffraction (XRD) 170
       2.1.1. Phase identification 170
       2.1.2. Solid solution identification 172
       2.1.3. Line Broadening Analysis (LBA) 172
   2.2. Small-Angle X-ray Scattering (SAXS) 173

2.3. X-ray absorption spectroscopy: EXAFS and XANES...................... 173

    2.3.1. Theoretical aspects 174

    2.3.2. Application of EXAFS to the study of metal catalysts.......... 175

    2.3.3. Application of XANES to the study of metal catalysts 175

    2.3.4. Conclusions 175

3. Electron microscopy 176

    3.1. Preparation of samples.. 176

    3.2. The different types of images and their use for the study
of metal catalysts... 176

    3.3. Energy Dispersive X-ray emission spectroscopy (EDX)................. 177

    3.4. Electron Energy Loss Spectroscopy (EELS) 178

    3.5. Conclusions... 178

LECTURE 10

Vibrational Spectroscopy with Neutrons

by H. Jobic

1. Introduction 181

2. Theory... 182

    2.1. Interaction of neutrons with matter................................. 182

    2.2. Vibrational spectroscopy with neutrons............................. 183

3. Experimental... 185

4. Examples 187

    4.1. Hydrogen chemisorption 187

        4.1.1. Hydrogen on nickel 187

        4.1.2. Hydrogen on ruthenium sulfide............................. 190

    4.2. Hydroxyl groups in HY zeolite...................................... 192

    4.3. Water in interaction with acidic sites in zeolites................. 196

    4.4. Benzene 198

5. Conclusion 199

LECTURE 11

Metal Catalysis in the Conversion of Biosustainable Resources

by P. Gallezot

1. Introduction 201
2. Hydrogenation reactions.. 203
   2.1. Hydrogenation of glucose into sorbitol 203
      2.1.1. Hydrogenation on Raney-nickel
         and supported nickel catalysts 203
      2.1.2. Hydrogenation on ruthenium catalysts.................................. 205
   2.2. Hydrogenation of triglycerides into edible oils and fats 206
   2.3. Hydrogenation of fatty esters into fatty alcohols............................. 207
3. Oxidation reactions... 208
   3.1. Oxidation of glucose to gluconic acid 209
   3.2. Oxidation of gluconic acid... 211
   3.3. Oxidation of 5-hydroxymethylfurfural 211
   3.4. Glycerol oxidation 212
   3.5. Future of metal catalyzed oxidations 213
4. Concluding remarks... 213

IR Characterization of Metal Catalysts using CO as Probe Molecule

F. Maugé, C. Binet and J.C. Lavalley

Laboratoire Catalyse et Spectrochimie, UMR 6506, ISMRA, Université de Caen,
6 boulevard du Maréchal Juin, 14050 Caen, France

1. Introduction

Infrared spectroscopy is a powerful technique to study the adsorption of molecules on well-defined metal surfaces or on metal supported on oxides. In particular for metal characterization, CO is frequently used as the probe molecule because of the high absorption coefficient of the $\nu(CO)$ mode and the sensitivity of its frequency to the oxidation state of the site, to its geometry and to its coordinative unsaturation degree. For spectra interpretation, it is important to take into account the dipolar coupling effects which induce wavenumber shifts and intensity transfer. In particular, caution should be taken in the characterization of defect sites. By following frequency shifts for various coverages, bandwidths and intensities, CO allows us to specify whether probe molecules form islands or are well-dispersed on the surface. Some examples of all these features are developed for single crystals and for supported metals.

2. Molecules adsorbed on a metal surface

2.1 Electric field

Considering a very thin film adsorbed on a highly reflecting metal surface, the wavelength of the infrared radiation (for $\nu = 2000$ cm^{-1}, $\lambda = 5$ µm) is much greater than the thickness of the adlayer (few angströms). Therefore, the adlayer thickness can be considered as infinitesimal compared to incident light wavelength.

The electric field of the incident radiation can be decomposed into two components : E_p and E_s (Fig. 1). The parallel component p presents an electric field vector in the plane of reflection. The perpendicular component is normal to the plane of reflection, i.e. parallel to the plane of the surface.

G. Maugé et al.

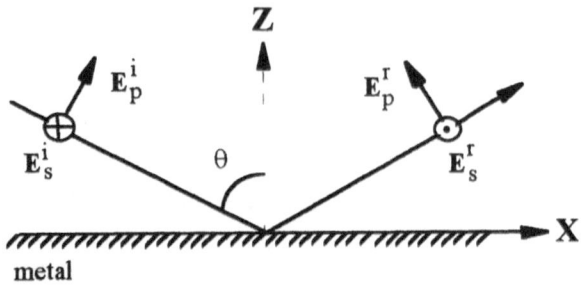

Figure 1 : Electric field vector for reflection of IR light on a surface.

From the work of Greenler [1], it is known that when light is reflected from a metal surface, the electric field vector of the incident radiation is subject to a phase change that depends both on its angle of incidence and on its polarization state (Fig. 2).

For the perpendicular component, the phase shift is close to 180° for all incident angles and the incident and reflected electric field vectors cancel. As for the parallel component, at small angles of incidence, the near zero phase shift induces a resultant electric vector close to zero. For $\theta = 90°$, the magnitude of the resulting electric field is also zero since the phase shifts are 180°.

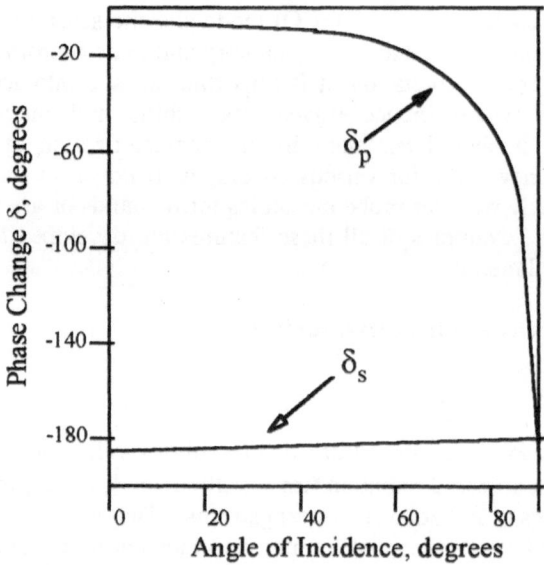

Figure 2 : Dependence of the phase change δ on reflection for the parallel (p) and perpendicular (s) polarizations of the electric field as a function of the incidence angle.

At large incidence angles, the vector sum of the incident and reflected electric vectors results in an electric field at the surface with a substantial component normal to the surface. As illustrated in Figure 3, the maximum of adsorption occurs at near grazing incidence.

In conclusion, the resultant electric field vector of the reflecting radiation has its main component pointing along the normal to the surface. Therefore, in order to excite a molecule adsorbed on a metal surface, incident light with a p-polarized component is required and only vibrations with a non zero dipole component normal to the surface will be active.

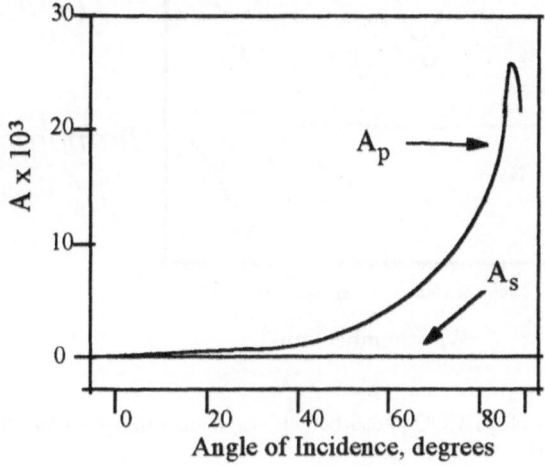

Figure 3 : Dependence of the absorption factors for the parallel (A_p) and perpendicular (A_s) polarizations at the wavenumber of maximum adlayer absorption as a function of angle of incidence. The calculation was for a 10-Å film of acetone on gold [2].

2.2 Selection rule : an application

An illustration of this selection rule was presented by Ito and Suëtaka [3] who studied the formation of formate ions by adsorption of formic acid on metal surfaces (Fig. 4). When the ions are adsorbed on a smooth copper surface, the reflection spectrum showed the 1360 cm^{-1} band corresponding to the symmetric OCO stretching vibration. By contrast, no band characterizing the antisymmetric OCO stretching vibration was observed near 1600 cm^{-1}. This absence is in agreement with the previous rule since the variation of the dipole moment for the latter vibration occurs parallel to the metal surface. Conversely, if the ions are adsorbed on a rough aluminum surface, both symmetric and antisymmetric OCO vibrations are active. This indicates that, on this non-planar surface, formate species are in various orientations.

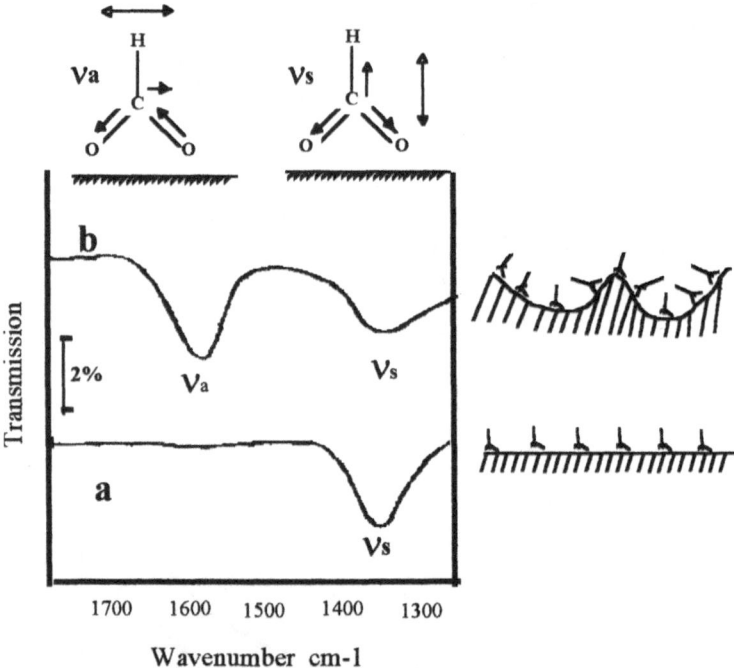

Figure 4 : IR spectra of [HCOO⁻] adsorbed on : (a) a smooth copper surface ; (b) a rough aluminum surface [3].

2.3 IR characterization of adsorbed molecules

On a monocrystal or a foil, the adsorbed molecules can be studied at grazing incidence by Infrared Reflection Absorption Spectroscopy (IRAS). The very high sensitivity of this technique, as previously developed, allows characterization of adsorbed molecules even for monolayer fractions. By using polarization modulation, the contribution of the gas phase can be separated from that of the adsorbed phase taking avantage of the selection rules for molecules adsorbed on a metal surface. Indeed, since only p-polarized radiation interacts vibrationnally with adsorbed molecules, by comparing spectra taken with p- and s- polarized light, it is possible to remove contribution arising from gas phase.

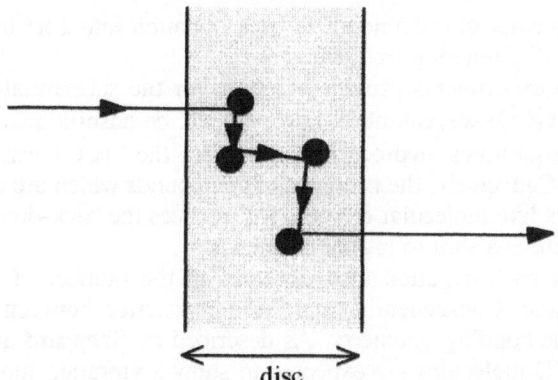

Figure 5 : Successive reflections on the metal particles of a supported catalyst.

On supported catalysts, spectra observed by transmission in fact correspond to successive reflections on the metal particles as illustrated on figure 5. Due to the multi-reflections within the disc, the angle and the polarization of the incident light vary for each reflection. Nevertheless, only the vibration presenting a dipole variation perpendicular to the surface of the metal particles will be excited.

3. CO Adsorption

3.1 Electronic interaction between the CO molecule and transition metals

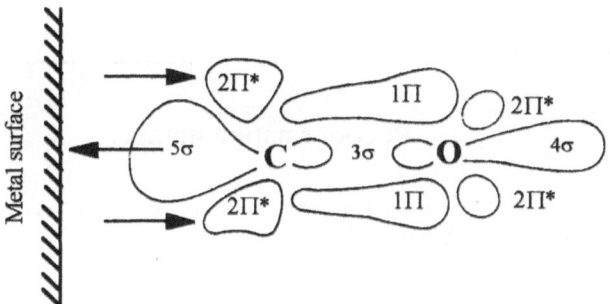

Figure 6 : Schematic representation of back-donation from a metal surface.

The interaction of the CO molecule with transition metals is described considering the frontier orbitals [15, 16]. The chemical bond between CO and metal involves electron donation from the 5 σ orbital of CO to the metal and a back-donation from metal d orbitals into the empty 2 π^* antibonding orbital of CO, as shown in figure 6. Since the 5 σ orbital is weakly bonding and the 2 π^* one is strongly antibonding, the C-O bond is weakened by the chemisorption process.

Therefore, the increase of the amount of back-donation into 2 π* orbital leads to a decrease of the C-O stretching frequency.

Co-adsorption experiments provide evidence for the substantial effect of back-donation on the ν(CO) wavenumber. For example, co-adsorption with an electron donor like an unsaturated hydrocarbon decreases the ν(CO) wavenumber up to 100 cm^{-1} [4, 5]. Conversely, the presence of compounds which are considered to be electron acceptors like molecular oxygen [6], reduces the back-donation to the CO molecule and induces a shift to higher frequencies.

The extend of back-donation also depends on the number of metal atoms to which CO is bound. Consequently, there is a correlation between the vibrational frequency and the bonding geometry. As described by Sheppard and Nguyen [7], linearly bound CO molecules are expected to show a vibration mode in the range 2130-2000 cm^{-1}, two-fold bridged in the range 2000-1880 cm^{-1} and three-fold bridged below 1880 cm^{-1}.

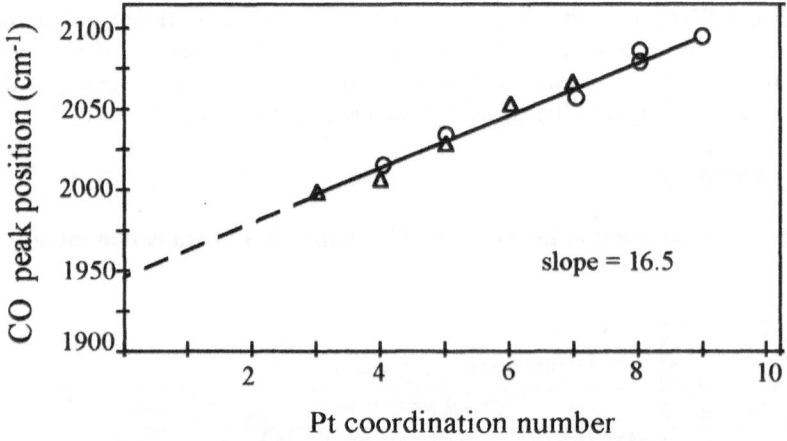

Figure 7 : Variation of ν(CO) wavenumber for lineary adsorbed species versus the coordination number of various Pt-catalyst [8].

More recently, it has been shown that the CO stretching frequency depends not only on the number of surface metal atoms to which CO is bonded but also on the coordination number of those atoms in the solid (Fig. 7). The metal coordination number decrease leads to lower ν(CO) frequencies. These results show that for on-top adsorption on low coordination sites, like edges and defects, CO can exhibit a wavenumber at the upper limit of the two-fold bridge absorption.

3.2 Frequency shifts for varying surface coverage

When the CO coverage increases, a shift to higher wavenumbers is generally observed. For a variation of the adsorbate density from an isolated molecule to a monolayer, it can reach 100 cm^{-1}. Two phenomena can give rise to this shift :
- vibrational coupling
- chemical shift.

3.2.1 Vibrational coupling (or dynamic effect)

In an adlayer containing species of similar vibrational frequency, a coupling phenomenon occurs. This effect called vibrational coupling or dynamic effect has been described extensively in the literature [9]. The oscillating molecules are assumed to interact via their through-space dipolar fields [10], through metal electrons, with their own image dipole and with the images of the other dipoles [11].

A consequence of this coupling is a shift to higher frequencies as coverage increases. Another less often reported effect of these lateral interactions is the reduction of the molecular absorption coefficient, ε_{CO}, as coverage increases [13, 14].

WITHOUT COUPLING   COUPLING INCLUDED

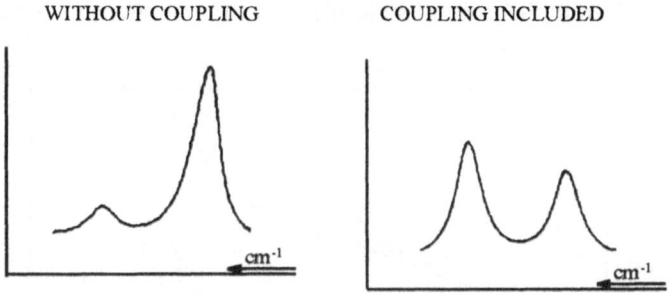

Figure 8 : Influence of dipolar coupling on the IR absorption [12].

When the adlayer presents more than one vibrational frequency, an intensity transfer can arise from the band due to the lower frequency vibration to that corresponding to the higher frequency vibration (figure 8). The magnitude of this effect is stronger for high adsorbate density, small frequency separation between the two vibrations and molecules absorbing strongly in the IR domain. In his review, Hollins [12] indicates that for carbon monoxide, coupling effects are small when the bands are more than 100 cm^{-1} apart, and negligible when the separation exceeds 200 cm^{-1}. Since linear and bridged-bonded carbon monoxide species generally exhibit wavenumber separations of at least 100 cm^{-1}, coupling between these vibrations is unlike to affect spectra very significantly. However, the separation between bands arising from adsorption on different coordination sites. like terraces and edges sites, are such that strong coupling effects can occur.

3.2.2 Chemical shift (or static effect)

The shift to higher frequencies observed as the coverage increases is also related to a modification of the CO-metal interaction. Blyholder [15, 16] suggested that increasing of the amount of adsorbed molecules induced a competition for back-donation electrons among these molecules. As a result, back-donation decreases and the frequency shifts to higher values.

3.2.3 Separation between dynamic and static effects

Except for column I B metal (see 3.2.4), both dynamic and static effects lead to an increase of vibrational frequencies as coverage increases. Nevertheless, it is possible to determine the contribution of each shift experimentally using isotopic mixtures. Indeed, dipolar coupling occurs only for neighbouring molecules oscillating at sufficiently close frequencies. At high dilution of ^{12}CO in ^{13}CO. ^{12}CO molecules are vibrationally decoupled while the chemical effect is unaffected. Therefore, comparison of the $\nu(^{12}CO)$ wavenumber due to pure ^{12}CO adsorption to the $\nu(^{12}CO)$ wavenumber due to ^{12}CO diluted in ^{13}CO for various coverages can be used to determine both effects. Thus, in the case of dilute ^{12}CO. the frequency shift from the isolated molecule to monolayer coverage corresponds

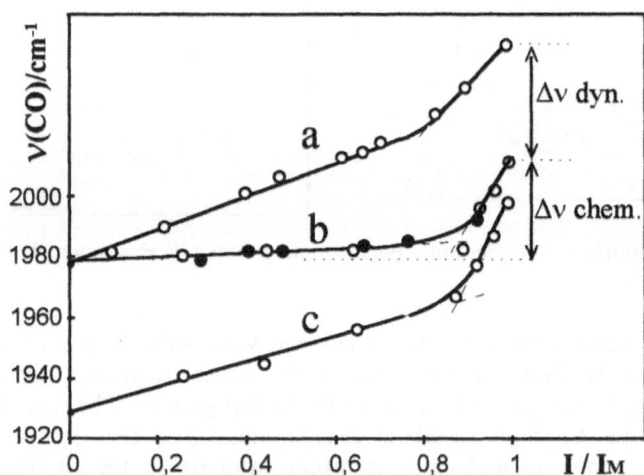

Figure 9 : Variation of the frequency as a function of coverage (I/I_M) of Ru/SiO_2 : (a) adsorption of ^{12}CO ; (b) and (c) adsorption of a mixture (10% ^{12}CO + 90% ^{13}CO). Curves (a) and (b) : $\nu(^{12}CO)$ variation. Curve (c) : $\nu(^{13}CO)$ variation [17].

to the chemical effect alone as opposed to the frequency shift measured for pure ^{12}CO which characterizes both chemical and vibrational effects. Consequently, the difference between these two shifts allows us to determine the vibrational shift value.

For example, on Ru supported on silica (Fig. 9) where only linear adsorption was observed, the total shift measured was 70 cm^{-1} [17]. From results obtained with (10% ^{12}CO + 90% ^{13}CO) isotopic mixture, this value has been decomposed into a shift of 40 cm^{-1} for the dynamic effect and 30 cm^{-1} for the static effect at saturation coverage. On a Pd(100) monocrystal, a two-fold bridged adsorption was detected [18]. At maximum coverage, around 35 cm^{-1} of the total shift (95 cm^{-1}) was ascribed to the dynamic effect, whereas around 60 cm^{-1} of the contribution corresponded to the static effect. Therefore, the shift due to the dynamic effect is unaffected by the nature of the substrate and the geometry of adsorbed CO species, as opposed to the static effect. As a matter of fact, there is more extensive back-donation from two-fold sites than from on-top site. This is in agreement with the greater wavenumber sensitivity of two-fold bridged species than linear species to the coverage variation.

3.2.4 Special case for column IB metals

Almost no frequency shift with coverage is observed for metals like Cu, Ag, Au. It has been shown for copper that, for increasing coverage, the chemical effect lowers the frequency, while the dipole effect raises it [19]. The balance between both effects leads to an almost constant wavenumber for the ν(CO) band.

3.3 Structure of the adsorbed molecule layer

In the adsorption process, competition occurs between diffusion and chemisorption of the molecule. The adsorption equilibrium can also be modified by the relative strength of adsorbed molecule - adsorbed molecule interactions and adsorbed molecule - metal interactions. These characteristics depend on the temperature of adsorption and on the coverage.

At very low coverage, if adsorbed molecule - adsorbed molecule attractives interactions dominate diffusion process, adsorbed species can form islands. Conversely, if the diffusion process dominates, adsorbed species form a disordered layer, called a lattice gas. For high coverage, the adsorbed layer structure can be developed bearing no relation to the metal lattice. But, if the adsorbed molecule - metal interactions are stronger, ordered structures of the adsorbed molecules are reported. In the case of the strongest interactions between the adsorbed molecule and the metal, a reconstruction of the surface can also be observed [20].

Considering a metal supported catalyst pressed into a disc form, if the chemisorption process dominates the diffusion process, the adsorption within the

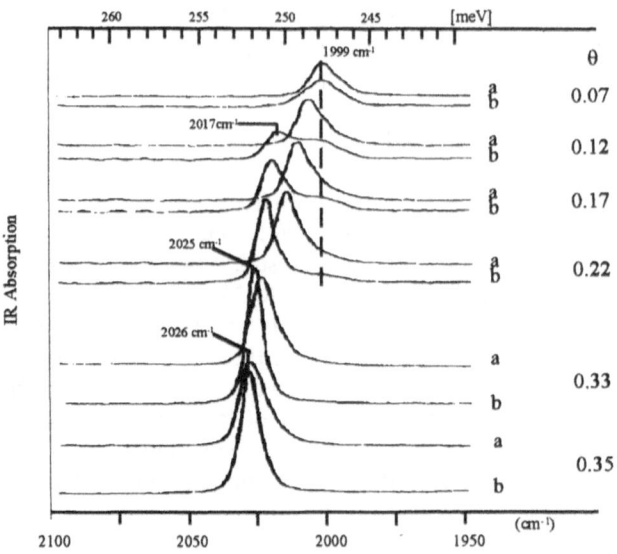

Figure 10 : Effect of annealing for CO/Ru(001) : **(a)** spectrum directly after adsorption at 80 K ; (b) spectrum after warming to 350 K and recooling to 80 K [18].

disc is inhomogeneous and first the edges of the disc are saturated by the adsorbed molecules

The high sensitivity of the ν(CO) wavenumber to the intermolecular distances and the variation of the ν(CO) bandwidth allow one to follow the homogeneity of the adsorbed layer with the coverage. Figure 10 illustrates the variation of the layer structure, versus the coverage and the temperature of CO adsorption on Ru(001) [18]. At 80 K, the low mobility of the molecule prevents island formation. By annealing at higher temperature (350 K) and re-cooling to 80 K, the band splits and two domains can be distinguished. The high frequency band can be ascribed to the formation of islands while the low frequency band characterizes a disordered phase. For low coverage ($\theta \sim 0.22$), increase of the CO coverage leads to increase island size at the expense of molecules in the disordered phase. At higher coverage, only one phase is present.

Variation of coverage can also modify the bonding mode. In monocrystal case, simultaneous characterization by LEED can allow the observation of well-defined adsorbed structure. On Pd (111), CO adsorption gives rise to a band which stays at a nearly constant frequency for coverage lower than 0.25 (Fig. 11). This indicates the formation of three-fold bridged species in a well-defined structure in agreement with simultaneous LEED observations which detect the presence of a ($\sqrt{3} \times \sqrt{3}$)-R30° unit cell. For higher coverages, the compression of the previous structure leads to the formation of a c(4 x 2) structure with all the molecules in two-fold sites as indicated by the large ν(CO) frequency shift.

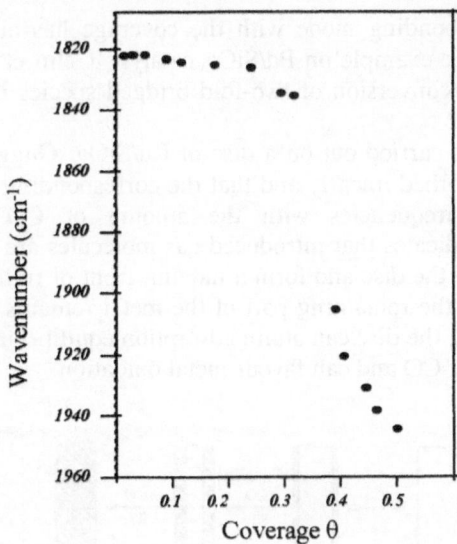

Figure 11 : Variation of the IR frequency as a function of coverage (θ) for CO/Pd(111) at 300 K [18].

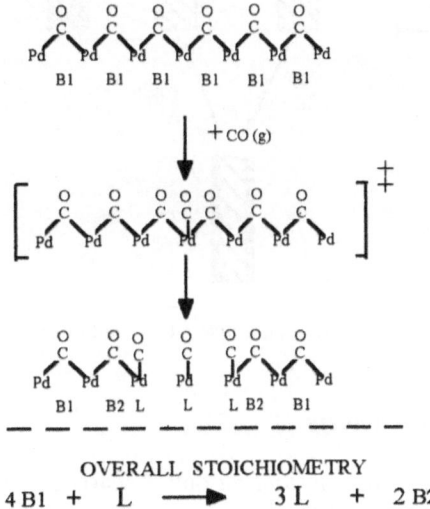

Figure 12 : Schematic one dimensional model for bridged-to-linear CO species conversion process [21].

Variation of the bonding mode with the coverage has also been reported on supported metals. For example on Pd/SiO_2 catalyst, Gelin et al. [21, 22] provided evidence for an interconversion of two-fold bridged species into linear species as described in figure 12.

From experiments carried out on a disc of Ru/SiO_2, Gugglieminotti et al. [17] show that CO is adsorbed linearly and that the corresponding spectra present only small changes in frequencies with the amount of CO adsorbed at room temperature. This indicates that introduced gas molecules are entirely adsorbed on the external layers of the disc and form a moving front of ruthenium particles near CO saturation while the remaining part of the metal remains free of adsorbed CO (Fig. 13). By heating, the disc can attain adsorption equilibrium but this can lead to partial dissociation of CO and can favour metal oxidation.

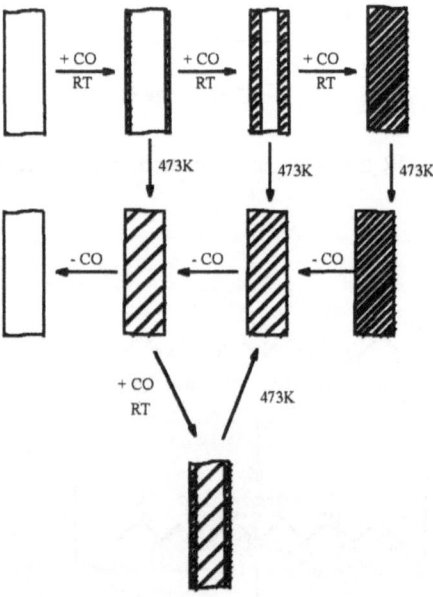

Figure 13 : Scheme of CO adsorption on a disc, at r.t. and after equilibration at 473 K [17].

3.4 Influence of surface defects

As presented previously and considering only linearly adsorbed CO species, it has been shown that the sites with the lowest coordination number such as edges and corners will present the lowest CO frequencies. The band characterizing terrace sites is situated approximately 20 cm^{-1} above that characterizing defects. However, due to this small wavenumber shift, dipolar coupling can occur between these two frequencies and leads to intensity transfer such that only the high wavenumber

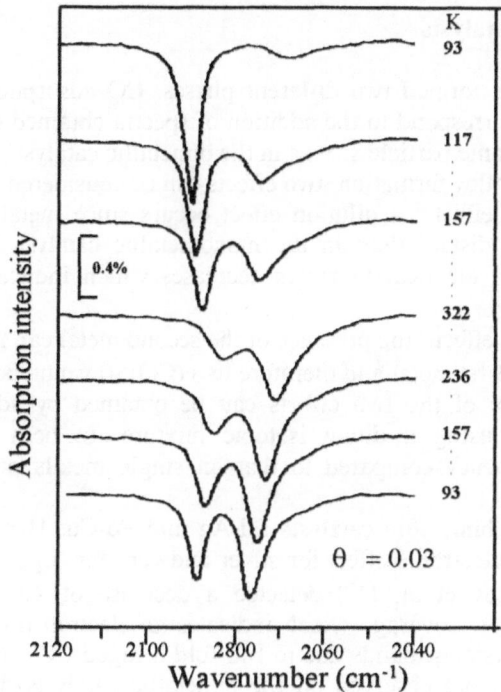

Figure 14 : Thermal evolution of low coverage CO/Pt(111) spectra [23].

band is detected. Therefore, spectra are dominated by the band characterizing terrace sites which does not reflect the proportion of such sites.

A way to control intensity transfer is to work at low coverage. For example, Tüshaus et al. [23] examined the influence of coverage on a nearly perfect Pt(111) crystal (Fig. 14). At very low coverage ($\theta = 0.03$) and for temperatures which allow CO diffusion (T $\geq$ 117K), two bands were detected at 2075 and 2090 cm^{-1} characterizing adsorption on defects and terraces, respectively. Conversely on this low defect concentration crystal, at higher coverage ($\theta = 0.2$), the defect band was no longer detected as a result of coupling to the terrace band. In a study of Pt(335), Xu et al. [24] observed adsorption on defects and terraces for $\theta_{CO} \approx 0.3$ but for lower coverage they detected only CO on defect sites while at higher coverage they observed only CO on terrace sites.

In the case of copper, Hollins [12] reported that the wavenumber associated with terrace sites is lower than that characterizing defect sites. Therefore, if intensity transfer occurs in that case, the band characterizing molecules adsorbed at step sites is the dominating one, while these species represent only a small proportion of adsorbed CO.

3.5 Bimetallic catalysts

If the two metals formed two different phases, CO adsorption on the bimetallic catalyst should correspond to the addition of spectra obtained on the monometallic catalysts with a same particle size as in the bimetallic catalyst.

In the case of alloy formation, two effects can be considered :

- a geometric effect : a dilution effect occurs since metal atoms of the same nature are more distant than in the monometallic catalyst. Therefore, dynamic coupling between identical vibrators decreases which induces a decrease of the $v(CO)$ wavenumber.

- an electronic effect : the presence of the second metal can modify the electronic properties of the first metal and therefore its $v(CO)$ wavenumber.

The separation of the two effects can be obtained by adsorbing CO at low coverage or by using a dilute isotopic mixture. In both cases, shift of the "singleton" frequency compared to that on single metals provides evidence for electronic effect.

On supported bimetallic catalysts Pd-Ag and Pd-Cu, Hendrix and Ponec [25] reported a weak electronic effect for silver and zero for copper. Conversely for Pt-Pb/Al$_2$O$_3$, Palazov et al. [26] detected a decrease of 40 cm^{-1} of the $v(CO)$ wavenumber at low coverage which indicates an electron transfer from Pb to Pt. On these three systems, bands due to two-fold bridged CO adsorption on Pd or Pt catalysts are no more observed on the bimetallic catalysts due to the geometric effect.

4. Examples of characterization of supported Pd catalyst

4.1 Evaluation of the relative amount of crystallographic faces on Pd/Al$_2$O$_3$

The IR characterization of the metal particles morphology was carried out on a series of Pd/Al$_2$O$_3$ (0.45% Pd) presenting dispersions in the range 9% - 51% [27]. In order to determine the saturation of the metal surface, area of IR bands characterizing metal adsorption were measured for CO increasing doses.

At saturation, the number of CO molecules introduced allowed us to calculated a metal dispersion assuming the stoichiometry CO/Pd$_{surface}$ = 1. It is interesting to note the good agreement between Pd dispersions determined by this method and by H$_2$ chimisorption.

At saturation, spectral profiles were strongly dependent on the dispersion (Fig. 15). The band attribution was proposed taking into account previous results obtained on monocrystals presenting various orientations [28, 29].

On supported catalysts, spectra provided evidence for CO adsorption on (100) faces (two-fold adsorption, band at 1965 cm^{-1}), on (111) faces (two and three-fold adsorptions, bands respectively at 1913 and 1834 cm^{-1}) and on terrace and edge sites (on-top adsorption, bands respectively at 2087 and 2062 cm^{-1}). As quoted previously, intensity transfer likely occurs between these two last bands which avoid any determination of the relative amount of these two sites. Therefore, the total area of these bands will be considered as a measurement of the discontinuity number. The $v(CO)$ integrated molar extinction coefficient was evaluated for a dense adsorbed layer, its value appeared to be independent of the CO adsorption structures. Therefore, it was possible to evaluate the relative importance of (100) faces, (111) faces and discontinuities for the variously dispersed palladium catalysts. Figure 16 provides evidence for a morphological change of the particles when their dispersion reaches 17% which corresponds to a particle diameter of 6.5 nm. The relative amount of the different faces showed that the large crystallites presented a spherical shape while the smaller crystallites developed a two-dimensional structure.

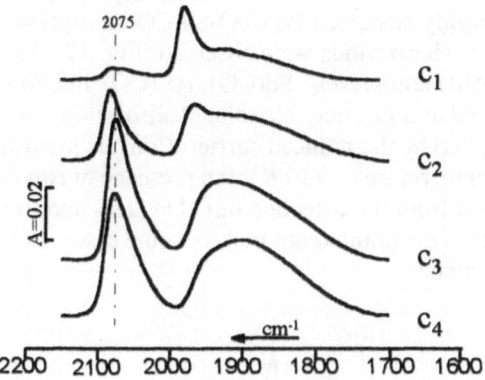

Figure 15 : Comparison of CO adsorption, for saturation dose, on Pd/Al$_2$O$_3$ reduced at 373K presenting various dispersions (C1 < C2 < C3 ~ C4) [27].

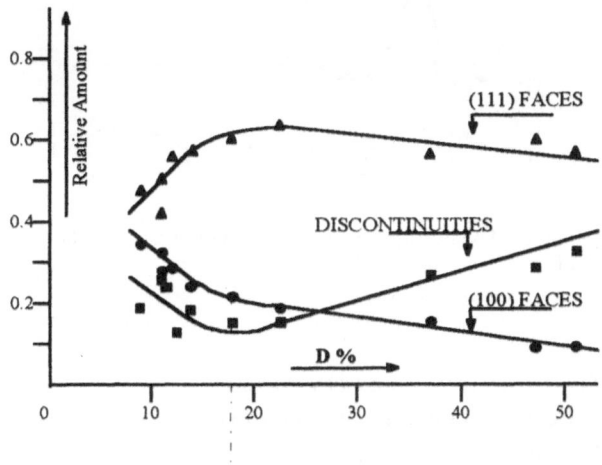

Figure 16 : Relative amount of faces and discontinuities versus the dispersion [27].

4.2 Formation of carbonyl clusters by CO adsorption on highly dispersed Pd/CeO₂

Supported metal particles can be present on the support as crystallites, as evidenced in the previous section, but some very small particles can also exist under the form of clusters, without well-defined crystallographic planes. Such structures were previously evidenced on metal encaged in zeolites [30]. The attribution of IR bands observed in this case, can be done by comparison with spectra obtained on carbonyl metal or colloidal metal [31].

Such clusters can also be formed by corrosive adsorption of CO on small crystallites [32]. For highly dispersed Pd/CeO₂, if CO adsorption is performed at room temperature, very sharp bands were observed (Fig. 17). They are relative to the formation of carbonyl complexes : $Pd(CO)$, $Pd_2(CO)$ and $Pd_2(CO)_2$. For these complexes, Pd and Pd₂ may be isolated adsorption sites either on small Pd agregates or stabilized by the reduced carrier. Conversely, when CO adsorption is performed at low temperature (~ 140 K), the presence of two dimensional crystallites was inferred from the detection of (111) faces and from the absence of detection of other sites. This pointed out that, in some cases, CO can modify the morphology of the particles.

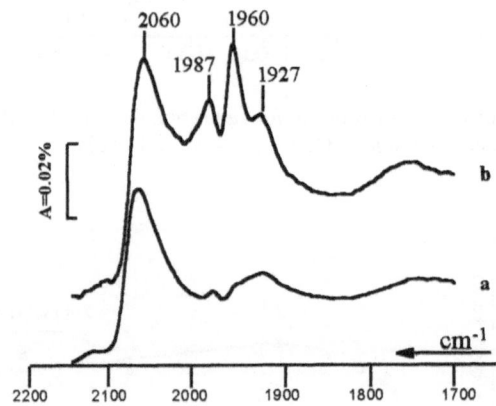

Figure 17 : Comparison of CO adsorption at low temperature (a) and at r.t.(b), on Pd/CeO₂ reduced at 623K [32].

REFERENCES

[1] Greenler R.G., *J. Chem. Phys.*, **44** (1966). 310.
[2] Golden W.G. Fourier Transform IR Spectroscopy, **Vol. 4**, Eds. J.R. Ferraro and L.L. Basile, Academic Press, (1985) p. 315.
[3] Ito M. and Süetaka W., *J. Phys. Chem.*, **79** (1975) 1190.
[4] Bertolini J.C., Dalmai-Imelik G. and Rousseau J., *Surf. Sci.*, **68**. (1977) 539.

[5] Ibach H. and Somorjai G., *Appl. Surf. Sci.*, **3** (1979) 293.

[6] Primet M., Basset J.M., Mathieu M.V. and Prettre M., *J. Catal.*, **29** (1973) 213.

[7] Sheppard N. and Nguyen T.T., *Adv. IR and Raman Spectrosc.*, **5** (1978) 67.

[8] Kappers M., Van der Maas J., *Catal. Lett.*, **10** (1991) 365.

[9] Hollins P. and Pritchard J., *Chem. Phys. Lett.*, **75** (1980) 378.

[10] Hammaker R.M., Francis S.A. and Eischens R.P., *Spectrochimica Acta*, **21** (1965) 1295.

[11] Mahan G.D. and Lucas A.A., *J. Chem. Phys.*, **68** (1978) 1344.

[12] Hollins P., *Surf. Sci. Rep.*, **16** (1992) 51.

[13] Hollins P. and Pritchard J., *Prog. Surf. Sci.*, **19** (1985) 275.

[14] Persson B.M.J. and Ryberg R., *Phys. Rev. B*, **24** (1981) 6954.

[15] Blyholder G., *J. Phys. Chem.*, **68** (1964) 2773.

[16] Blyholder G., *J. Phys.Chem.*, **79** (1975) 756.

[17] Guglielminotti E., Spoto G. and Zecchina A., *Surf. Sci.*, **161** (1985) 202.

[18] Hoffmann F.M., *Surf. Sci. Rep.*, **3** (1983) 107.

[19] Hollins P., Davies K.D. and Pritchard J., *Surf. Sci.*, **138** (1984) 75.

[20] Somorjai G. and Van Hove M., *Progress Surf. Sci.*, **30** (1989) 201.

[21] Gelin P. and Yates J., *Surf. Sci.*, **136** (1984) L1-8.

[22] Gelin P., Siedle A. and Yates J., *J. Phys. Chem.*, **88** (1984) 2978.

[23] Tüshaus M., Schweizer E., Hollins P. and Bradshaw A.M., *J. Elect. Spect. and Rel. Phen.*, **44** (1987) 305.

[24] Xu J., Henriksen P.N. and Yates J.T., *Langmuir*, **10** (1994) 3663.

[25] Hendrickx H.A.C.M. and Ponec V,. *Surf. Sci.*, **192** (1987) 234.

[26] Palazov A., Bonev Ch., Kadinov G. and Shopov D., *J. Catal.*, **71** (1981) 1.

[27] Binet C., Jadi A. and Lavalley J.C., *J. Chim. Phys.*, **86** (1989) 451.

[28] Bradshaw A.M. and Hoffmann F.M., *Surf. Sci.*, **72** (1978) 513.

[29] Greenler R., Burch K., Kretzschmar K., Klauser R., Bradshaw A., Hayden B, *Surf. Sci.*, **152-3** (1985) 338.

[30] Sheu L., Knozinger H. and Sachtler W., *J. Mol. Catal.*, **57** (1989) 61.

[31] Bradley J.S., Millar J.M., Hill E.W., Behal S., Chaudret B. and Duteil A., *Faraday Discussion*, **92** (1991) 255.

[32] Badri A., Binet C. and Lavalley J.C., *J. Chim. Phys.*, **92** (1995) 1333.

Preparation of Industrial Catalysts

Ch. Travers

Institut Français du Pétrole, B.P. 311, 92506 Rueil-Malmaison cedex, France

1. INTRODUCTION

Catalysts are used in all important industrial reactions and particularly in refining and petrochemistry. By definition a catalyst is a product which allows to accelerate a chemical reaction, and remains unchanged during this reaction. It plays a role on the kinetic of the reaction, decreasing the activation energy by creation of several energetical intermediate states. Catalysts can be homogeneous: solved in a liquid phase, or heterogeneous usually in the solid state. These heterogeneous catalysts are either bulk catalysts, only made up by active species, or supported catalysts, in which the active species are dispersed on a preliminary shaped support. In this paper we will focus on the heterogeneous supported catalysts, which present the advantages to use less active species especially when this species is expensive, to obtain a better dispersion of this latter on the support and to take advantage of the textural properties of the support.

2. CATALYSTS PREPARATION

2.1. Generalities

Preparation of industrial catalysts begins with the choice of active phases and supports, based on literature, previous experience, and of course fundamental research, always taking in mind that to be competitive, this catalysts must be prepared at the lowest possible cost.

The most often used supports in catalysis are:

• transition alumina: obtained by thermal treatment of hydrated alumina. The nature of the final transition alumina depends on the structure and texture of the hydrated precursors, and on the thermal treatment conditions. The most commonly used alumina in catalysis are gamma and eta alumina, which respectively develop specific area of 150-200 sqm/g and 250-300 sqm/g, and Lewis acidity corresponding to about 5 OH/sqnm.

• silica, amorphous solid which can develop specific area as high as 800 sqm/g, and a rather low Brönsted acidity.

• amorphous silica-alumina, with different alumina content. The L.A. (low alumina) type contains about 15 wt% of alumina and the HA (high alumina) type contains about 25 wt% of alumina. Their specific area are comprised between 200 and 400 sqm/g, and their acidity is essentially of the Brönsted type.

• zeolites or crystallised silica-alumina, which form a very important part of the acidic supports, used in a wide range in catalysis and adsorption for their properties of cation exchangers, molecular sieves, and for their adjustable acidity by the mean of chemical treatments. The acidity is essentially of Brönsted type.

The choice of the active phase depends on the reaction which has to be catalysed.

The principal active phases can be under metallic, oxide or sulphide forms and are presented in Table I.

Even if catalysts preparation seems relatively easy for the non-specialist, it is a rather complex field, constituted by numerous elementary steps, each governed by physical phenomena as liquid diffusion, mass transfer in porous medium etc.

Each step depends on several parameters as ph, temperature, duration, pressure, etc., which must be carefully controlled, because each of them can have a fundamental influence on the final product.

From the precursor to the final catalyst an important number of elementary steps are handled like precipitation, impregnation, thermal and hydrothermal treatments (drying, calcination, reduction), separation, washing, mixing, grinding, shaping and all the special treatments directly performed in the industrial unit: chlorination, sulfuration, regeneration. Even if some of them appear as minor like washing, mixing they have to be also carefully understood and controlled, because they can also influence strongly the characteristics of the final catalyst. Before reviewing the principal elementary steps, we have to remind some essential basic laws as Young-Laplace, Kelvin and Ostwald laws.

2.2. Theoretical notions

We have to consider two important laws, involved in several elementary steps as impregnation, thermal treatment...

Table I.

Active agents	Examples	Reactions
1 - Metals	Fe, Co, Ni,Ru, Rh, Pd, Ir, Pt, Ag, Cu, Zn	Hydrogenation Dehydrogenation Hydrogenolysis Cyclisations Oxydations-NH
2 - Sulphurs and oxides	NiO, CuO, ZnO, CoO, Cr_2O_3, V_2O_5, MoO_3... WS_2, MoS_2, Ni_3S_2, Co_9S_8 MgO, La_2O_3	Oxidations Reductions Dehydrogenation Cyclisation Hydrogenation Desulfuration and desazotation
3 - Acidic and non-conductors oxides	H-Zeolites SiO_2-Al_2O_3 SiO_2-MgO Al_2O_3 + X Supported Friedel-Crafts	Hydratations and deshydratation Isomerisations Polymerisations Alkylations Cracking

2.2.1. Young-Laplace law

It expresses the pressure difference between two phases separated by a curve interface, the highest pressure being located in the concave side of this interface.

In the case of Figure 1 we have

$$\Delta P = P - P' = \gamma \left[\frac{1}{r_1} - \frac{1}{r_2} \right] \qquad (1)$$

where γ is the surface tension.

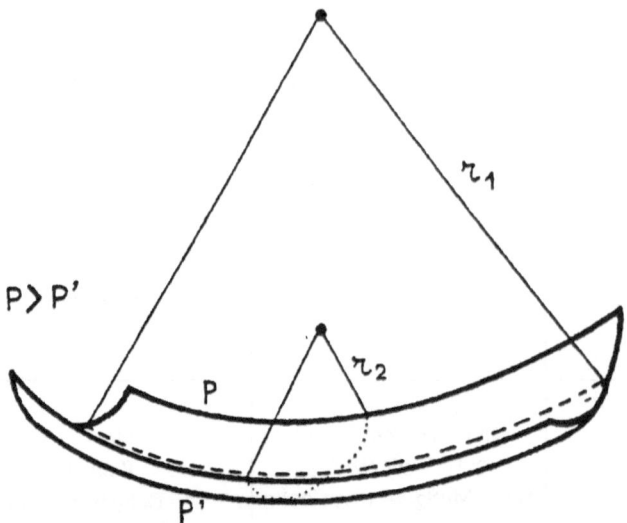

$P \rangle P'$

Fig. 1 — Young-Laplace law.

If the interface is spherical, the relation (1) becomes

$$\Delta P = P - P' = \frac{2\gamma}{r} \text{ with } r = \text{curve radius}$$

this is the case of a drop of water.

2.2.2. Kelvin and Ostwald laws

We only consider the most current case of a spherical interface. The vapour pressure of a liquid separated from its gas phase by a curve interface is different of the vapour pressure of this same liquid separated from its gas phase by a plane area.

$$RT \text{ Log } \frac{p}{p\infty} = \pm \frac{2\gamma V_M}{r}$$

γ = surface tension
V_M = molar volume of the liquid
r = radius of the curve
p = vapour pressure of the liquid separated of its gas phase by a curve interface
$p\infty$ = vapour pressure of the liquid separated of its gaseous phase by a plane interface

We use + if the center of the curve is in the liquid phase: case of a drop of water.

We use – if the center of the curve is in the gas phase: case of water in a capillary tube (Fig. 2).

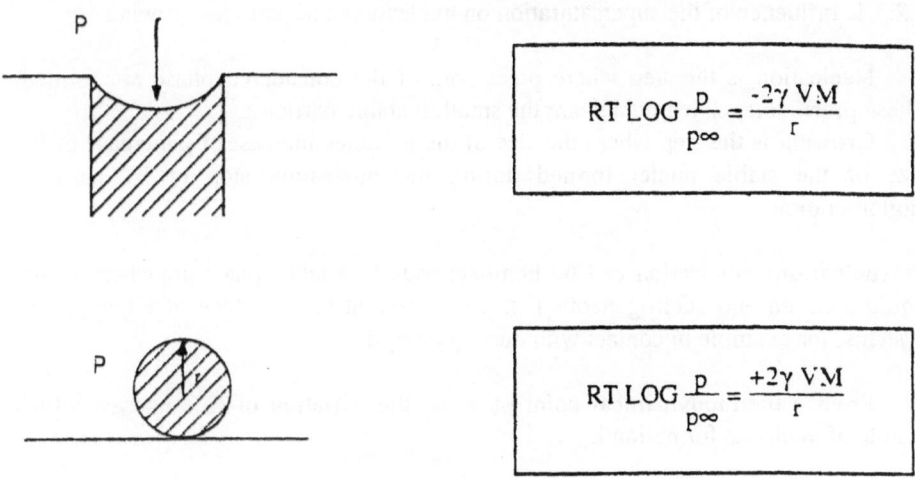

$$RT \, LOG \, \frac{P}{p\infty} = \frac{-2\gamma \, VM}{r}$$

$$RT \, LOG \, \frac{P}{p\infty} = \frac{+2\gamma \, VM}{r}$$

Fig. 2.

This law has been adapted to the solutions by Ostwald

$$RT \, Log \, \frac{C}{C_\infty} = \frac{2\gamma \, V_M}{r}$$

with C = concentration of a solution in equilibrium with small crystals with a radius equal to r; C∞ = concentration of a solution in equilibrium with crystals with an infinite radius.

Simply, these laws indicate that:
• smaller is the drop, more it tends to evaporate (Kelvin law),
• smaller is the particle, more it tends to solve (Ostwald law),
and this is not compatible with the highly dispersed character of a catalyst. Here we have to introduce the concept of supersaturation.

2.2.3. Supersaturation. Nucleation and growing of particles

A catalyst containing one or several dispersed phases can only be prepared starting from a reactionnal medium with high free energy, that means with a strong supersaturation.

The supersaturation is one of the fundamental parameters which govern the catalysts preparation.

2.2.3.1. Influence of the supersaturation on nucleation and particles growing

Nucleation is the step where precursors of the considered phase are formed. These precursors, called nuclei, are the smallest stable particles.

Growing is the step where the size of the particles increases by increase of the size of the stable nuclei formed during the nucleation step or by particles agglomeration.

a) Nucleation: Nucleation can be homogeneous if it takes place anywhere in the liquid medium and heterogeneous if it takes place at the interface of a two-phases systems, for example in contact with dust, crystals defect etc.

From a thermodynamical point of view, the variation of free energy, which results of nucleous formation is

$$\Delta G^\circ = N \left[\mu_2 - \mu_1\right] + \gamma S$$

With μ_2, μ_1:chemical potentials of the considered compound respectively in the solution and in the nucleous.

N: number of atoms or molecules in the nucleous
S: nucleous area
γ: surface tension
$N \left[\mu_2 - \mu_1\right]$ is a negative volumic term, proportional to R^3 (R=nucleous radius), which represents the variation of free energy during the transformation of the compound from the solution state to the nucleous.
γS is a positive surface term proportional to R^2, which represents the energy we have to give to the system to increase the interface from zero to S.

If now we represent the curve $\Delta G = f(R)$ (Fig. 3), we notice that:
• it is positive and growing for the small R values
• it goes through a maximum ΔG^* for a R^* value, radius of the critical nucleous, the most instable particle of the system;
• it goes through zero for a R_s value, radius of the stable nucleous, the smallest stable particle of the system. With mathematical demonstrations (2), which are not the aim of this paper we have access to the expression of ΔG^*, R^*, N^* for the critical nucleous and R_S, N_S for the stable nucleous.

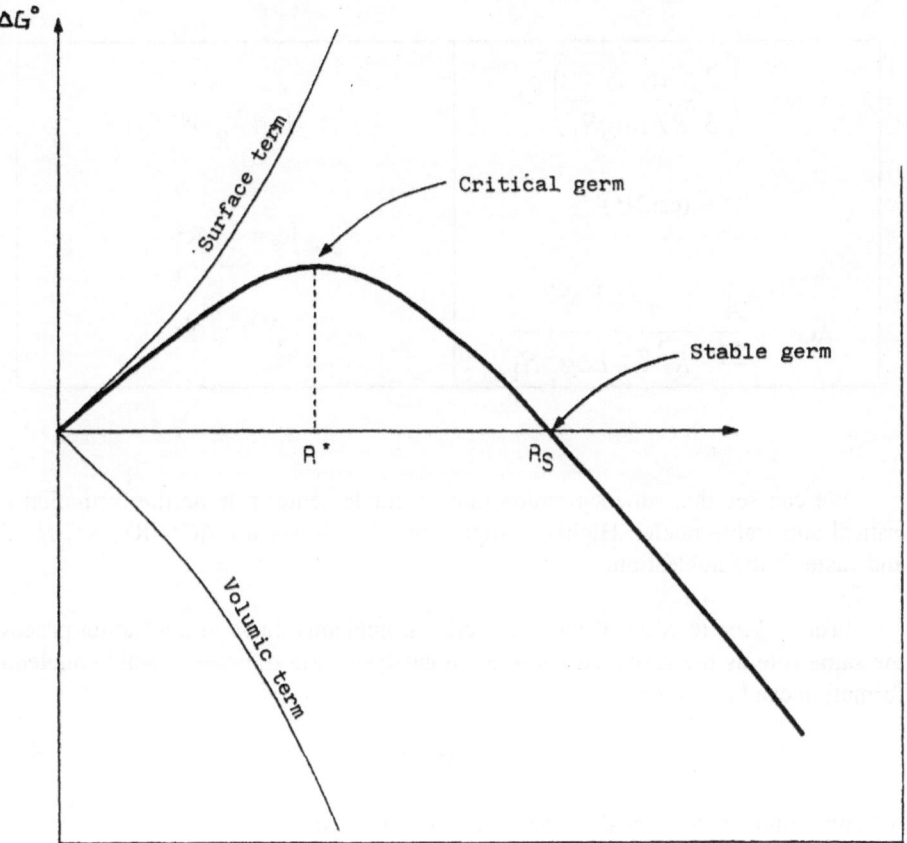

Fig. 3.

Assuming that $\mu_2 - \mu_1 = \dfrac{RT}{N} \, \text{Log} \, \dfrac{C_\infty}{C_1}$

where C_∞ = equilibrium concentration for particles with $R = \infty$; $C1$ = effective concentration of the particles in the solution; N = Avogadro number; we can define the supersaturation as $S_1 = \dfrac{C_1}{C_\infty}$, and give the expression of R^*, N^* R_S, N_S as a function of this supersaturation (Table II).

Table II.

$$N^* = \left(\frac{2}{3} \frac{a\gamma N}{RTLogS_1}\right)^3$$

$$N_s = \frac{27}{8} N^*$$

$$R^* = (cst.N^*)^{1/3}$$

$$R_s = \frac{3}{2} R^*$$

$$\Delta G^* = \frac{4}{27} \frac{a^3\gamma^3 N^2}{R^2 T^2 Log^2 S_1}$$

$$\Delta G_s^* = 0$$

We can see that supersaturation plays a fundamental role on the formation of critical and stable nuclei. Higher is supersaturation, lower are ΔG^*, R^*, N^*, N_s, R_s and easier is the nucleation.

From a kinetic point of view, the critical nucleous plays in nucleation process the same role as the activated complex in catalysis, and the rate of stable nucleous formation can be written:

$$Vg = n^* \, S^* \, V$$

with n^* = number of critical surface nuclei/volume unit

$$n^* = ne^{\dfrac{-\Delta G^*}{RT}}$$

and V = fixation rate per surface unit of solved species on the critical nucleous

$$Vg = A \exp\left(-\frac{4}{27} \bullet \frac{a^3\gamma^3 N^2}{R^3 T^3 Log^2 S_1}\right)$$

The formation rate of stable nucleous is a function of the supersaturation too.

At low supersaturation Vg is negligible.

When S_1 is higher than a critical value called threshold of homogeneous rate (S_{hm}) the nucleation rate strongly increases.

In the case of heterogeneous nucleation $S_{ht} \ll S_{hm}$ (Fig. 4).

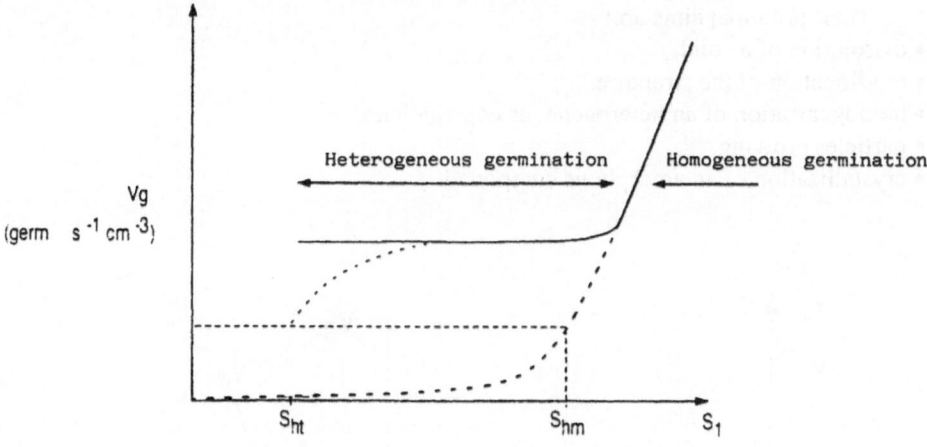

Fig. 4.

b) Growing: Growing of particles can occur by increase in size or by coalescence. Whatever the mechanism, the equation rate can be written

$V_c = k (S_1 - 1)^2$ for low supersaturation
$V_c = k' (S_1 - 1)$ for high supersaturation

Refering to Figure 5 where V_g and V_c evolution is drawn versus supersaturation, we can see that when:

• at low supersaturation, $V_c > V_g$: formation of big crystals is favoured.
• at high supersaturation, $V_g > V_c$: formation of small crystals is favoured .

2.3. Review of the main elementary steps

Among the numerous elementary steps involved in the preparation of industrial catalysts, we will focus on
• thermal or hydrothermal treatments
• forming
• metal impregnation

2.3.1. Thermal or hydrothermal treatments

Hydrothermal treatments allow to transform a solid in the presence of water at a temperature comprises between room temperature and about 500 °C. It can be performed under pressure and is more particularly devoted to gel and precipitate.

Their principal aims are:
• dissolution of a solid,
• modification of the structure,
• homogenisation of an heterogeneous coprecipitate,
• particles growing,
• crystallization of an amorphous compound.

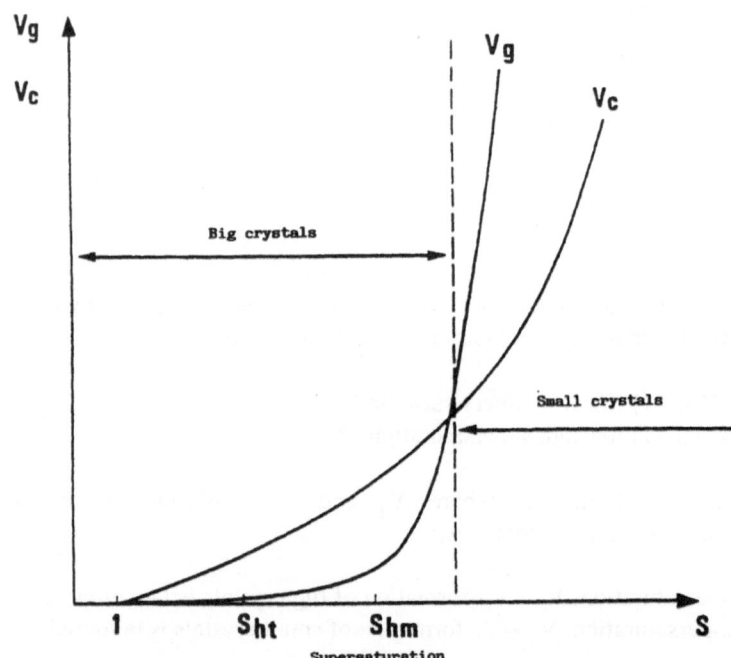

Fig. 5.

Thermal treatment is the operation during which the solid is heated at a temperature higher than 200 °C under controlled atmosphere. The principal thermal treatments are drying and calcination.

Their principal aims are:
• elimination of volatils compounds,
• obtaining will defined structure and texture for support and active agent,
• giving strong mechanical strength,
• regeneration of the catalyst.
    Among the various types of chemical or physico-chemical transformations that occur during calcination the following are the most important:

• creation of a generally macroporous texture through decomposition and volatilization of substances previously added to the solid at the moment of its shaping,

• modifications of texture through sintering: small crystals or particles will turn into bigger ones,

• modifications of structure through sintering, as for example, the transformation of alumina crystals according to:

$$Al_2O_3 \text{ } \eta \text{ cubic} \rightarrow \theta \text{ monoclinic} \rightarrow \alpha \text{ hexagonal}$$

• thermal decomposition reactions leading to active agents or to the precursors of active agents, as well as to gaseous products creating texture and to textural and structural reorganization of the solid products of decomposition,

• thermal synthesis reactions with or without elimination of volatile compounds, generally followed by modification of structure and consequently of texture in the synthesized products.

All the calcining transformations obey the laws of thermodynamics. The decompositions and the syntheses with elimination of volatile products are complete only, if the kinetics are favourable. The variations of texture will be oriented toward the formation of small surfaces; and crystalline structures will only be obtained from amorphous phases if the exothermal heat of transformation is enough to compensate for the reduction in entropy that goes with the reorganization of the system, the same as for hydrothermal syntheses.

As with most of the reactions related to the chemistry of solids, the reactions that take place during calcination are very complex, and great care must be taken in the study of their kinetics, particularly when the speeds frequently depend on the presence of impurities that have been added, intentionally or not. Modifications of texture and structure are limited by phenomena of transfer of matter occurring either within the mass of the particle, on its surface, or by passage into the gaseous phase. It is known that these transfers, which are the driving forces of sintering, become important only in the vicinity of the Tamman temperature, which is defined as half of the absolute temperature of fusion of the solid. Surface diffusions will first appear at lower temperatures (T_{tamman} x 0.7); but since they are less energetic than the diffusions in the mass, they will be supplanted by the mass diffusions at a higher temperature (T_{tamman}) while diffusions via the gaseous phase will only appear later still. Sintering, corresponds to the textural and structural transformations that a solid undergoes when it is submitted to a thermal treatment.

2.3.2. Forming

In the case of supported catalysts, forming usually precedes calcination step, and impregnation of the active agent.

Two extreme types of forming can be distinguished, depending on whether the desired product is powder microgranules or granules on the order of one to several millimeters.

2.3.2.1. Forming microgranules

Crushing and grinding often serve only to prepare a charge for forming into granules. As a general rule, crushing and grinding is done by generating successive shocks between the product to be crushed and a very hard mass making up the grinder or crusher. The necessary kinetic energy is either furnished by the product, as in cyclone-pulverizers, or more usually by the equipment, as in ball-crushers and mixer-grinders. The lower limit of size for the powder is a few microns.

Spray-drying, accomplishes forming and drying at the same time. It consists of spraying microdroplets of the product to be dried into a hot gas current. It is suitable only for making small-diameter beads (7-700 μm), because the violence of the treatment provokes local superheating in the solids, leading to difficulty in releasing steam that is suddenly produced, with consequent mechanical degradation of large beads.

Spray drying is used for obtaining microbeads used in fluidized beds.

Drop coagulation results from metastable sols suspended in a different liquid phase, and can simultaneously achieve gelling, ripening and forming. The aqueous sol is distributed in the form of droplets by a sparger whose orifices are sized to give the desired diameter of bead. The droplets settle through the water-immiscible solvent, whose temperature is raised to around 100°C. The surface tension created on the droplets during passage through the solvent permits formation of gel spheres that must be ripened and ultimately dried. During drying, a contraction of the bead without deterioration is observed as long as the diameter is not too large. This treatment is suitable for producing either microbeads or beads on the order of a few millimeters.

2.3.2.2. Forming granules

The raw material comprises a calibrated powder or paste with suitable rheological properties. No matter what method is to be used, the charge should show properties of fluidity and adhesiveness.

The different usual methods are pelletizing, extrusion and pan granulation.

A. Pelletizing: This consists of compressing a certain volume of usually dry powder in a die between two moving punchers, one of which also serves to eject the formed pellet. Fluidity of the powder is required to assure homogeneous filling of the die; a certain amount of plasticity is desirable in the granules to create the maximum contact between them; and the quality of the intergranular contacts will depend, after compression, on the adhesive properties of the powder.

If a powder does not have all the required qualities, one can add lubricants to help the sliding and positioning of the microgranules. Such lubricants can be liquid (water, mineral oil) or solid (talc, graphite, stearic acid, and various stearates). Binders are also added to increase the post-compression adhesion, as for example starch is added for pelletizing active carbon. An increase in the adhesive forces due to chemical bonds resulting from contact between the granules can also be achieved by peptizing the microgranules, whose surfaces are thus made more chemically reactive.

Some of the operating variables are characteristic of the equipmen, while others depend on the charge and, for a given powder, on the granulometry, the kind and concentration of binders, and the lubricants. The ratio of the diameter of the powder particles to that of the final granules should fall between 1/20 and 1/50.

A conventional industrial pelletizer equipped with around thirty dies can produce 5-10 liter/h of pellets a few millimeters in diameter. Such pellets are usually cylindrical with flat or rounded base surfaces (better distribution of the compression forces); however spheres, hollow cylinders, and toroids can also be obtained.

B. Extrusion: Extrusion is a rather general technique applied to pastes; one device forces the paste through a die, while another cuts off the extruded material at the desired length. The ease of extrusion and quality of the product depend on the following properties of the paste:

a) Viscosity: A non-thixotropic product that is too viscous will block the extruder. A product that lacks in viscosity can not be extruded with a screw and will give extrudates without mechanical resistance when extruded by a press.

b) Thixotropy: Certain substances become less viscous under shearing forces, and then recover their initial state after the forces have been released for a time called the relaxation time. The existence of such thixotropic properties is eminently favorable for the flow of a paste and formation of a solid granule at the exit of a die, providing the relaxation time is short enough.

c) Stability: Under extrusion conditions, there should be no dynamic sedimentation of the product through exuding water and forming a paste that is too viscous.

d) Homogeneity: The paste must be homogeneous to assure that the quality of the product is constant. When necessary, the paste is homogenized in a mixer-kneader under controlled conditions of temperature, time, and pH. An excess of kneading can in fact compact the material and suppress potential macropores. Screw extruders partially knead the paste as it travels along the screw.

Extruding equipment can be classed in one of two categories: press extruders. and screw extruders. Press extruders are used principally for pastes that are viscous; screw extruders are preferred for thixotropic products.

Even for a given charge with specific properties, the operating variables are related to the type of equipment. Generally they include: temperature, addition of binders and lubricants to modify viscosity and thixotropy, as for example additions of alginates, starch, kaolinite and montmorillonite.

Extrusion granules generally occur as cylinders 0.5-10.0 mm in diameter. Hollow cylinders can also be obtained with special dies. We have to notice that conditions of extrusion sometimes lend themselves to certain hydrothermal transformations, and one must be very carefull, and remember that extrusion is really a part of the catalyst preparation.

c) Pan granulation: This consists of agglomerating a powder into beads by moistening it as it rolls about in a rotating pan. Seed granules are coated with humidified powder under the effect of capillary tensions to give a bead of increasing diameter, by a kind of snow-ball phenomenon. The discharge of the beads in the desired size is accomplished through centrifuging. The granulation operation is followed by a ripening period that starts in the bowl. To achieve good granulation the powder must be rather fine (dp < 50 μm). For a powder with given internal properties the operating variables are: speed of rotation, inclination of the pan, rate of flow of water, and nature and quantity of the added binders.

Pan granulation is not expensive but has the drawback that the product has a rather wide size distribution, necessitating a screening operation afterwards. The diameter of the beads can vary from 1 mm to 20 mm.

The choice between the differents forming method depends on the starting material as shown in Table III.

Table III. — Influence of starting material on the choice of a grain-forming process.

Starting material	Process
Monolithic blocks	Crushing and grinding
Powders	Pelletizing Granulation Extrusion
Pastes, hydrogels	Extrusion
Sols	Drop coagulation
Melted solids	Spray-drying

and on the process in which the final catalyst will work. For example, in continuous processes, the catalyst will be imperatively under beads form to avoid attrition problems. In fixed bed process, both beads and extrudates can be used.

2.3.3. Impregnation of active agents on a support

This elementary step consists in dispersing the active agent on the support which may be inert or catalytically active.

The active agent is never introduced into a porous support in its final form but by the intermediary of a precursor, the choice of which holds great importance for the quality of the final deposit, its structure, its grain size, its distribution as a function of the diameter of the granule. Two types of impregnating can be considered, depending on if an interaction exists between the support and the precursors at the moment of wetting, or if there is no interaction.

The first one is the more complex type of impregnation and comprises, ionic exchange, grafting, hydrolytic precipitation. We will only describe here the ionic exchange.

2.3.3.1. Impregnation without interaction

It exists two types of impregnation without interaction:
• capillary impregnation
• diffusional impregnation

a) Capillary impregnation: It consists in wetting the dried support developping a pore volume VPT with a volume V of solution, containing the precursor of the active phase, with the right concentration. The solution goes through the support porosity. Usually V = VPT.

This type of impregnation is characterised by:
• a strong exothermicity, usually without great importance, but which must be nevertheless carefully controlled, in the case of particular supports like zeolites for example;
• high pressures developed in the porosity: during the impregnation, air bubles are confined in the porosity. The pressure in these bubles depends on r, radius of the spherical interface liquid-gas and follow the Young-Laplace law (cf II.2.1.)

$$\Delta P = P - P' = \frac{2\gamma}{r}$$

with $P' = 10^5$ Pa
$\gamma = 7.10\text{-}2$ N.m^{-1} for water

if   R = 50 nm, P = 29 b
if   R = 2 nm, P ~ 700 b
      Smaller will be r, higher will be the pressures in the bubles and higher will be the risk of grains breaking.

Impregnation under vacuum is one of the considered solutions to prevent this phenonemon.

• <u>duration of the impregnation</u> which can be determined connecting Poiseuille and Young-Laplace laws for a pore with a radius r

$$Q = \frac{V}{t} = \frac{\pi r^4}{8\eta} \times \frac{\Delta P}{L} (1) \text{ with } V = \pi r^2 L$$

Q = flow rate in the pore of volume V
η = dynamic viscosity (Pa.s)
L = length of the pore

ΔP = pressure drop = $\dfrac{2\gamma}{r}$ (2)

$$(1) + (2) \rightarrow \boxed{t = \frac{4\eta L^2}{\gamma r}} \text{ time required to fill the pore of volume V}$$

if   L = 1 mm
$\gamma = 7.10^{-2}$ Nm^{-1}
$\eta = 10^{-3}$ Nm^{-2}.s
t ~ 1 s for a pore of radius r = 50 nm
t ~ 30 s for a pore of radius r = 2 nm
t ~ 60 s for a pore of radius r = 1 nm

The filling of the pores is very fast.

Industrially two main processes are used (Fig. 6):
• a discontinuous one by aspersion
• a continuous one by immersion

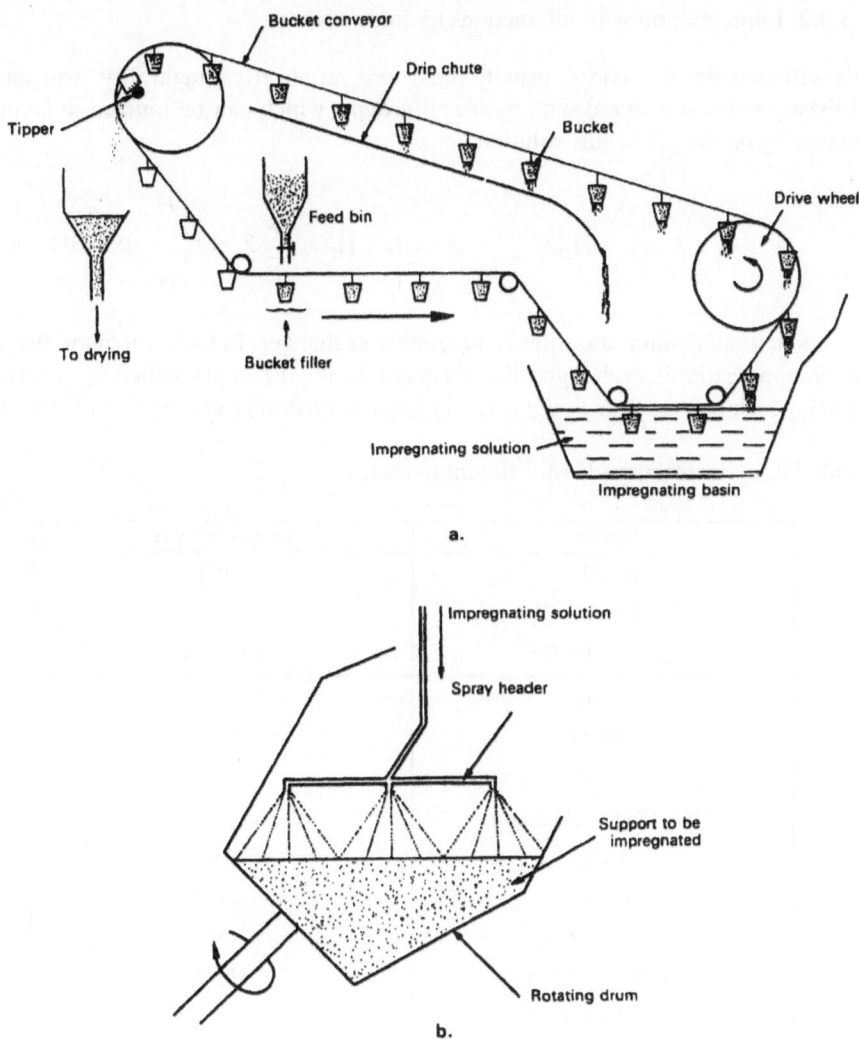

Fig. 6. — Two types of industrial support-impregnating process.

b) Diffusionnal interaction: For this type of impregnation, the porosity of the support is firstly filled with the solvent of the impregnation solution. Then the saturated support is immersed in the impregnation solution.

This type of impregnation doesn't undergo exothermicity, and high pressures in the porosity, but the migration time is of course longer than in the case of capillary impregnation, that's why this method is practically never industrially used.

2.3.3.2. Impregnation with interactions by ionic exchange

We will consider the oxides, usually presenting an electrical neutrality. The surface of these oxides is covered with hydroxyl groups, which can be ionized in solution, depending on the pH of this solution:

$$
Z\text{-}O^- + H_3O^\oplus \; \underset{2}{\overset{1}{\rightleftarrows}} \; Z\text{-}OH + H_2O \; \underset{2}{\overset{1}{\rightleftarrows}} \; Z\text{-}O\Big\langle {}^{H}_{H} \;\; \oplus + OH^-
$$

In acidic medium, the solid is an anionic exchanger. In basic medium, the solid becomes an cationic exchanger. For every solids, it exists a pH, called isoelectric pH or pH_i which corresponds to the perfect electrical neutrality of the solid (Table IV).

Table IV. — Isoelectric pH of different oxides.

Oxides	Isoelectric pH
Sb_2O_5	< 0.4
Hydrated WO_3	< 0.5
Hydrated SiO_2	1.0 - 2.0
U_3O_8	~ 4
MnO_2	3.9-4.5
SnO_2	~5.5
TiO_2 (Rutile, Anatase)	~6
UO_2	5.7-6.7
γFe_2O_3	6.5-6.9
Hydrated ZrO_2	~6.7
Hydrated CeO_3	~6.75
Hydrated Cr_2O_3	6.5-7.5
$\alpha, \gamma Al_2O_3$	7.0-9.0
Hydrated Y_2O_3	~8.9
αFe_2O_3	7.4-9.0
ZnO	8.7-9.7
Hydrated La_2O_3	~10.4
MgO	12.1-12.7

At pH < pH_i, the solid will be an anionic exchanger.
At pH > pH_i, the solid will be a cationic exchanger. It is really essential to know the pH_i of the support before proceeding to the impregnation.

But one must be very careful, when the solutions are strongly basic or acidic, because dissolution of the support can occur (Fig. 7).
We will consider successively cationic and anionic exchanges.

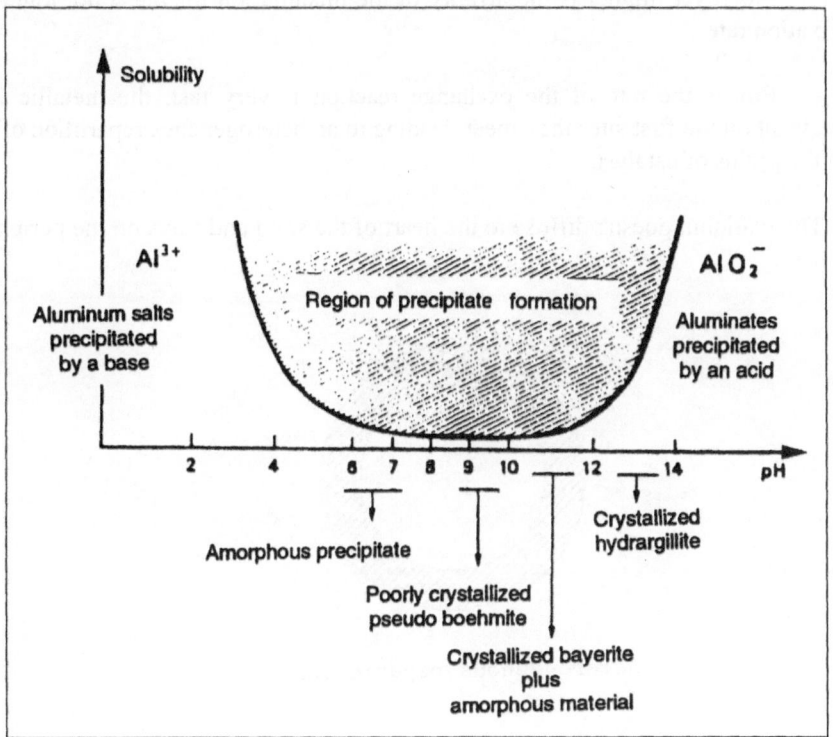

Fig. 7.

- **Cationic exchange:** For example, we want to introduce 0.3 wt% of platinum, in a zeolite under H-form, zeolites being natural cationic exchangers. The following reasoning could be made to deposit platinum on a support at pH > pH$_i$.

  First of all, we have to choose, the soluble platinum salt able to give a cation. $Pt(NH_3)_4Cl_2$ seems to be a good one.

  The equilibrium between H$^+$ ions of the zeolite and Pt $(NH_3)_4^{2+}$ of the solution can be written:

$$Pt\ (NH_3)_4^{2+}\ (solution) + 2H^+\ (zeolite) \underset{\leftarrow}{\rightarrow} Pt\ (NH_3)_4^{2+}\ (zeolite) + 2H^+$$
$$(solution)$$

It can be demonstrated (3) that lower is the amount of engaged platinum compared to the number of free sites in the zeolite, higher is the fixation rate of the metal.

Moreover, higher is the affinity of the metallic ion for the solid, higher is the fixation rate.

But as the rate of the exchange reaction is very fast, the metallic ions strongly sit on the first sites they meet, leading to an heterogeneous repartition of the Pt in the grains of catalyst.

The platinum doesn't diffuse to the heart of the solid and stays on the periphery (Fig. 8).

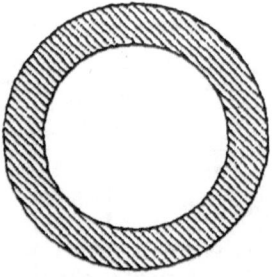

Heterogeneous repartition

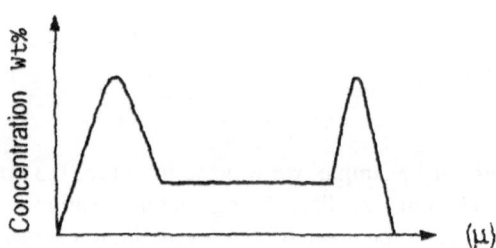

Fig. 8. — Macroscopic repartition of metal during simple ionic exchange.

To develop a high activity, usually all the metallic atoms must be accessible to the reactant, that means that the Pt repartition, must be homogeneous in the volume of the grain.

To reach this good repartition, it is necessary to add a component able to decrease the diffusion rate of the Pt to the surface of the grain. This result is obtained adding a cation, call a competitor ion, which enter in competition with $Pt(NH_3)_4^{2+}$. Usually we choose NH_4^+.

$$NH_4^+ \text{ (solution)} + H^+ \text{(zeolite)} \underset{\leftarrow}{\rightarrow} NH_4^+ \text{ (zeolite)} + H^+ \text{(solution)}$$

The competition ratio is defined as $R = \dfrac{NH_4^+}{2Pt(NH_3)_4^{2+}}$ mol/mol.

Using this technic of exchange with competition, we get an homogeneous repartition for the metal (Fig. 9).

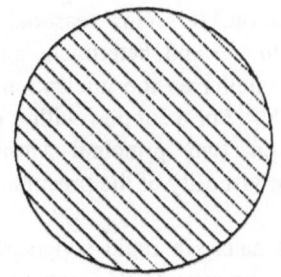

Homogeneous repartition

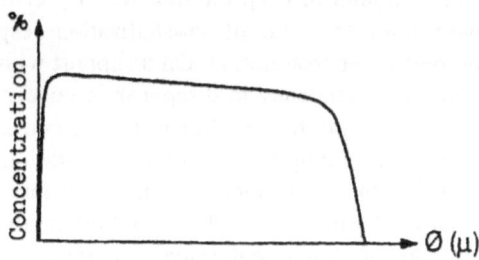

Fig. 9. — Macroscopic repartition of metal with competitiv ionic exchange.

In case of anionic exchange, if pH < pH_i, we have to choose a salt able to give an anion, for example H_2PtCl_6, the hexachloroplatinic acid.

The exchange and competition reactions are then:

$$(H_2 Pt Cl_6)^{2-} \text{ (solution)} + 2OH^- \text{ (support)} \underset{\longleftarrow}{\longrightarrow} (H_2 Pt Cl_6)^{2-} \text{ (support)} + 2OH^-$$
$$\text{(solution)}$$

$$Cl^- \text{ (solution)} + OH^- \text{ (support)} \underset{\longleftarrow}{\longrightarrow} Cl^- \text{ (support)} + OH^- \text{ (solution)}$$

2.3.4. Unit operations after impregnation

The main operations are drying and calcining.

<u>In the case of impregnation without interaction</u>, drying consists in causing the precursor to crystallize in the pores of the support through supersaturating the solution, generally by means of simple evaporation. Crystallisation within a porous support is in principle analogous to crystallization of bulk catalysts during precipitation, except for the important difference due to the high surface contact, between the support and the solution. It seems reasonable to expect that this environment would be favorable to a rapid heterogeneous nucleation, which is desirable because the crystals are smaller as the nucleation rate, V_N, is increased relative to the speed of nucleation, V_G. For a support with a given surface, the ratio of these two speeds depends on two principal variables: the degree of supersaturation (see II.2.3.1) and the viscosity of the precursor solution.

Rapid vaporization of solvent causes a rapid supersaturation of the residual solution, and thus tends to increase the ratio V_N/V_G, and to favor a deposit of fine crystals. However this rapid vaporization can be effective only as long as the solution's viscosity is high enough to keep the transfer of precursor solute into the residual solution slower than the rate of crystallization, especially when the precursor solution is far below supersaturation. On a support with one narrow range of pore sizes, the solvent has a tendency to disappear in concentric layers starting from the outside of the granule, and this might cause an accumulation of precursor toward the heart of the granule as long as the rate of evaporation is slow in relation to the rate of diffusion of the solute. In supports with a wide range of pore sizes, the biggest pores will first empty themselves of their solution, so that there is a risk of selective accumulation of precursor within the smaller pores.

This drying is followed by calcining whose aim is to fix the chemical structure of the precursor prior to final activation (reduction, sulfurization) in the commercial reactor just before start-up. Also, calcining can adjust the surface and texture of the deposit so as to obtain optimum catalytic properties.

With supported catalysts, it is necessary to take into account possible interactions between the support and the precursor of the active agent, which interactions are likely to occur at temperatures of calcination. If the decomposition of the precursor is highly exothermic and liberates a large volume of volatile products (i.e., acetates, citrates and oxalates) a high concentration of the precursor frequently causes intense bursting of the support granules; it is therefore preferable to carry out the calcination in two stages, a first stage of slow calcining with a low oxygen content in the gases and a second stage conventional calcining.

In the case of impregnation with interaction, drying conditions rarely influence the qualities of the catalyst. Nevertheless, it is adviced to carefully control this step too.

Drying conditions rarely influence the qualities of the catalyst; the recommended temperatures are generally too low for decomposing the precursor.

Calcining after ion exchange is governed by the same operating variables as those already discussed for impregnations without interaction. The sintering of the active phase is generally slowed down because of the extreme dispersion of the precursor; however, this does not prevent conditions of temperature and gas composition etc., from having a considerable influence on the final state of the catalyst.

Depending on whether or not the catalyst is reduced directly after drying or treated with air before reduction, the size of the platinum crystallites varies considerably.
To optimize the dispersion of the metallic phase it is generally recommended to calcine before reduction and other further treatments.

Finally, it should be noted that impregnation by ion exchange leads to a much better dispersion of the active agent than does simple wetting without exchange. The metal dispersion resulting from impregnating by ion exchange remains excellent even when the amount of metal deposited on the support reaches several weight percent, whereas wetting the support leads to the formation of large crystallites whose size depends on the deposited amount of platinum and which do not consequently afford a surface proportional to the platinum concentration (Fig. 10).

At this stage of preparation, the catalyst is ready to be loaded in the industrial unit, in which it will be submit to other elementary steps as, reduction, sulfurization... if necessary.

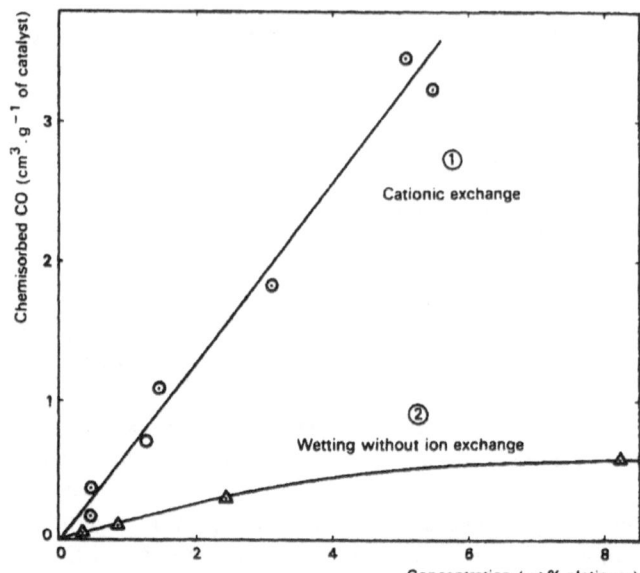

(1) Ion exchange from a
solution of Pt(NH$_3$)$_4$Cl$_2$.
(2) Wetting with a solution of
H$_2$PtCl$_6$.

Fig. 10. — A comparison between two methods for impregnating platinum on a silica support.

3. CONCLUSION: EXAMPLE OF PREPARATION OF INDUSTRIAL CATALYSTS

We describe (Table V) the preparation of the industrial zeolitic catalyst for light paraffins isomerisation starting from a sodium mordenite with a Si/Al (atom) ratio equal to 5.

This preparation consists of at least 8 elementary steps, each of them of great importance and with a well defined aim.

It is the combination of all these steps which conditions the properties of the final catalyst. When preparing a catalyst, we must keep in mind that each step (even the more simple), each parameter of this step is fundamental and must be carefully controlled. Nothing must be let to the chance. This is the key of successful processes.

Table V. — Elementary steps for preparing the zeolitic industrial catalyst for light paraffins isomerisation.

Na mordenite	Aim
Ion exchange with $NH_4 NO_3$	to replace Na^+ by NH_4^+
Hydrothermal treatment: steaming	to adjust the Si/Al ratio to a higher value
Acid leaching	
Shaping	to make extrudates
Thermal treatment (drying + calcining)	to eliminate volatils products and give high mechanical strength to the support
Metal impregnation by competitive exchange	to deposit the active agent
Thermal treatment (drying + calcining)	to decompose the metallic precursor and generate the oxide
Thermal treatment (reduction)	to generate the active metallic species

This course is mainly drawn from

(1) J.F. Le Page, Applied heterogeneous catalysts, Technip Edition

(2) Ch. Marcilly, IFP Internal report n° 30077

(3) Ch. Marcilly, Journal of the Algerian Chemical Society, acts 2eme Coll. Franco-Maghreb. Catal. side Fredj, 1-5/06/92, tome 1 p. 34-98.

Measurement of Catalyst Performances at the Laboratory

M. Forissier

Laboratoire de Génie des Procédés Catalytiques, CPE Lyon, 43 boulevard du 11 Novembre 1918, B.P. 2077, 69616 Villeurbanne, France

Summary.
　　This paper may help researcher in catalysis to develop catalyst testing and kinetic study at the laboratory. The following points are discussed :
- Some definitions: simple and complex reactions, elementary steps, kinetic expressions, reaction rate, conversion, selectivity, residence time, time on stream.
- Diffusion limitation, external and internal diffusion, experimental evidence.
- Experimental reactor to study catalytic reaction in the laboratory: batch reactor, flow reactor, differential flow reactor, perfectly mixed flow reactor, pulse reactor.
- Reaction scheme.
- Kinetic model.
- Catalyst deactivation.
- Model selection and kinetic constant determination.
- Interpretation of kinetic constants values.
- Mechanism of reaction.

1. CATALYST TESTING AND KINETIC STUDY

1.1. Catalyst testing
　　A catalytic reaction is performed in normalised conditions, closed to industrial conditions. It is generally impossible to keep exactly the same reaction conditions.
　　Global measurements are made : conversion, selectivity... a very good reliability is expected.
The limitations:
　　No data on the reaction mechanism
　　For different catalysts, the optimum results of a catalyst are generally not obtained at the same reaction conditions. The normalised test do not give the best results of a catalyst.

1.2. Kinetic study

It is the measurement of reaction rates, in a large field of experimental conditions, when the rates are controlled by chemical phenomena,

- to find the reaction scheme or the list of chemical reactions involved,
- to observe the effect of partial pressure or concentration of reactants and products on the rates of each reaction,
- to propose for each reaction one (or more) list of elementary steps able to explain the observations,
- to make one (or more) mathematical kinetic model able to simulate the observations,
- to select the best model, to obtain the values of its kinetic constants, with the precision on the constants,
- to verify experimentally the consequences of the model,
- to interpret the values of constants by the chemistry, to build a detailed mechanism of the reactions,
- to simulate by the calculation the production of a supposed industrial reactor, introducing limitations due to mass and heat transfers and to hydrodynamic phenomena.

This method is not very often used because it seems a too long way to applications. Nevertheless it give a very complete knowledge of the catalytic reaction system and permits a good optimisation of the industrial unit and a fast adaptation to the economical conditions (productivity, purity of products, evolution of the catalysts, ...)

2. SOME DEFINITIONS

2.1. Complex chemical reactions

There is apparently no stoichiometric ratio between reactants and products. Several simple chemical reactions occur one after the other or in parallel. A simple chemical reaction shows constant stoichiometric ratio.

2.2. Elementary step

A simple stoichiometric reaction is the result of a succession of elementary steps. Each of them involves one or tow reactants which may be :
- a gaseous molecule and a catalytic site,
- an adsorbed molecule,
- an adsorbed molecule and a gaseous molecule,
- two adsorbed molecules,
- an adsorbed molecule and a site,
- one or two gaseous molecule (non catalytic step).

Often one of these steps, slower than the others, impose its rate to the simple stoichiometric reaction.

The elementary steps may be direct or direct and inverse or equilibrium. In this last case the rate of the direct and the inverse reactions are large compared to the rate of the simple reaction.

2.3. Kinetic expression

Only the rate of elementary step may be theoretically predicted. The rate results from molecular collisions between molecules or sites, it is proportional to reactant concentration (or partial pressure) and to a rate constant.

For the elementary step : $A + B \rightarrow C$

$$r = \frac{dn_C}{dt} = kC_A C_B \text{ or } r = \frac{dn_C}{dt} = k' P_A P_B$$

The constants k and k' are only function of the temperature and of the nature and quantity of the catalyst

If the steps involve the site X of the catalyst, the coverage ratio θ is used

$$\theta = \frac{\text{number of sites ready to work in the step}}{\text{number of sites able to work in the step}}$$

For instance :

$A + X \rightarrow AX$ ads $\qquad\qquad r = kP_A \theta_X$

2.4. Reaction rates

Various rates are defined (some formula to obtain rates are given for flow reactor).

Absolute rate	r	Molar flow rate of a product without regard for the catalyst	$r = v . x$
Rate by volume unit	r_V	Molar flow rate of a product for a volume unit of catalyst bed	$r_V = r / V$
Volume by volume by hour		Flow rate of product in m^3 / h	
	VVH	by reactor volume in m^3.	
Specific rate	r_m	Molar flow rate of a product by mass unit of the catalyst	$r_m = r / m$
		or of active phase	$r'_m = r / m'$
Rate by area unit	r_i	Molar flow rate of a product by area unit of the catalyst	$r_i = r / m . s$
		or of active phase	$r'_i = r / m' . s'$
Turn over number	N	Molar flow rate of a product by 1 site-gram or 1 atom-gram of active metal (working frequency of the site).	$N = r.M / m'.D$

with :

v (mol. s-1), Total molar flow rate at the outlet of the reactor;

x, molar ratio of the product at the outlet of the reactor;

V, volume of the catalyst bed;

m, mass of catalyst;

m', mass of active phase in the reactor;

s, specific area of the catalyst ;

s', specific area of the active phase;

M, atomic or molar mass of the active species;

D, active phase dispersion : it is the ratio of the atom number of the active phase at the catalyst surface to the total atom number of the active phase .

2.5. Conversion, yield and selectivity

The conversion X (or "taux de transformation global" TTG, in French) is, for the reactant (r) :

$$TTG_r = X_r = \frac{\text{mole nomber of r that reacted}}{\text{mole nomber of r that entered into the reactor}}$$

The yield to a product (p) Xp (or "taux de transformation en un produit" TTp is :

$$TT_p = X_p = \frac{\text{mole nomber of r transformed to p}}{\text{mole nomber of r that entered into the reactor}}$$

Remark : $X_r = \sum_p X_p$

The selectivity Sp to a product (p) is :

$$S_p = \frac{\text{mole nomber of p obtained}}{\text{maximum mole nomber of p that may be theorically obtained}}$$

It is necessary to give the reaction that we hope to realised, with its stoichiometric coefficients v_p.

$$Sp = TTp / TTGr = Xp / Xr \text{ et} : \sum_p S_p = 1$$

2.6. Time on stream and residence time

These tow times must be distinguish in flow reactor. The time on stream is the duration between the beginning of the work of the catalyst and the time of the measurement.

The residence time is the time used by a molecule to go from the inlet to the outlet of the catalyst bed. This time is generally different for each molecule and a residence time distribution is considered.

- a mean residence time is : $$t_{sm} = \int_{inlet}^{outlet} \frac{dV}{Q}$$

dV, an elementary volume in the reactor,
Q, volume flow rate through the elemental volume.

- the contact time: $tc = V / Qo$
V, reactor volume;
Qo, flow rate through the reactor.
Remark : if the stoichiometric coefficient of the product $v_p = 1$, then :

$tc = 1 / VVH$ (en hour)

3. CHEMICAL OR DIFFUSION LIMITATIONS

These points are well reported in books, for instance: [1 to 5].

The external diffusion limitation appears when the reactant consumption at the external surface of the catalyst is so quick that an important concentration gradient take place between the solid surface and the fluid phase. It is probable when the rate by catalyst surface unit are large, and when the fluid velocity is small..

The internal diffusion limitation appears when concentration gradients are important along the pores. It is probable with porous catalyst, large pellets diameters, large reaction rates..

The estimation of effectiveness coefficients by correlation [4] is not always reliable and it is often useful to control it by some experiments

3.1. In a differential flow reactor

A flow reactor is differential (see § 423) if partial pressures or concentrations of reactants are practically the same between the inlet and the outlet of the reactor. In this case the conversion is proportional to the catalyst mass and to the residence time if there is no limitation by external diffusion.

Nevertheless if the catalyst temperature is not homogenous (exothermic effect) a limitation by diffusion may be hidden.

If the conversion is proportional to the catalyst mass, and to $1/D$ (D is the total flow rate, or if the conversion is independent of the ratio m/D, it may be conclude that the reactor is a differential reactor and that the external diffusion is not limiting.

Moreover if the reaction rate do not depend of the pellet diameter, and if the Arrhénius plot is a straight line and if the apparent activation energy is not too small (>15kcal), the internal diffusion is not limiting, and the chemical limitations are observed.

3.2. In an integral flow reactor

With an integral reactor it is not possible to do the same tests
This reactor is perfect to hide a diffusion effect:

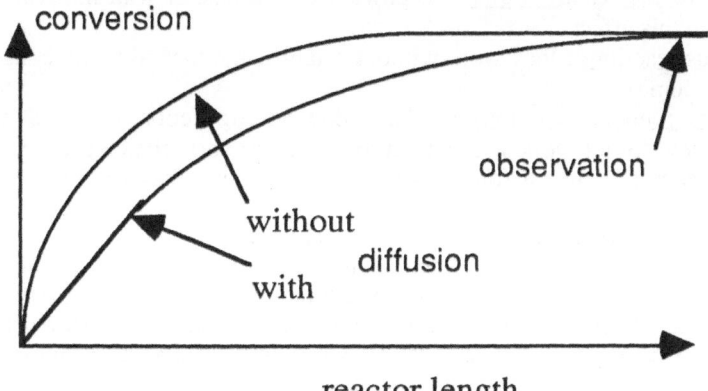

If the chemical kinetics is known, it is possible to compute the theoretical conversion at the outlet of the reactor and to compare with the experiments.

4. EXPERIMENTAL DEVICES FOR KINETIC STUDIES

4.1. Global point of view
The chemical analysis of reactants and products is very important for the kinetic study. One of the rare methods to verify the chemical analysis is to make mass and atomic balances between the inlet and the outlet of the reactor.

4.1.1. Choice of the experimental conditions of the study
- Reactor design: it is useful to control experimental conditions to simplify calculations.
- The present phases : choice the contact conditions
- Reactor dimensions : limits of the analysis sensibility, measurements of reaction conditions, reliability of the catalyst sample...
- Pellet or powder dimensions : small pellets or powders are good to eliminate internal diffusion limitations but introduce large pressure-drops in tubular reactors.
- Catalysts mass, reactant flow rate, dilution. Do not forgot analysis and security problems.

4.1.2. The control and the knowledge of the reaction conditions
It is very important to know :
- the temperature of the catalyst bed (the temperature regulator only know the sensor temperature),
- the partial pressures of reactants,
- the speed of the fluid flow,
- the evolution of the catalyst properties (deactivation, fouling, sintering,...)

4.1.3. Some experimental difficulties to detect
- Catalytic effect of the reactor walls: use powder of the reactor wall material as catalyst to estimate this effect.
- The homogenous reactions : they work without catalyst, and increase with contact time (or reactor volume).
- The hetero-homogeneous reactions (at least one homogeneous step and one catalytic step) : they do not work without catalyst and are very difficult to detect. The effect of the reactor volume before and after the catalyst bed may be used.

4.1.4. How to obtain practically the chosen reaction conditions
- A catalyst bed with a large exchange area, catalyst dilution with an inert powder, dilution of the reactants are used to obtain a constant temperature in the catalyst bed.
- Flow rate regulators or good vapour saturators are useful to control inlet flow rate,
- The complete analyse of product is necessary to verify mass balance at each run.

4.2. Some laboratory reactors often used

4.2.1. The batch reactor

a) Principle.
Reactants are put in contact at time 0. The product formation and reactant disappearance are measured during time.

b) Some practical designs.

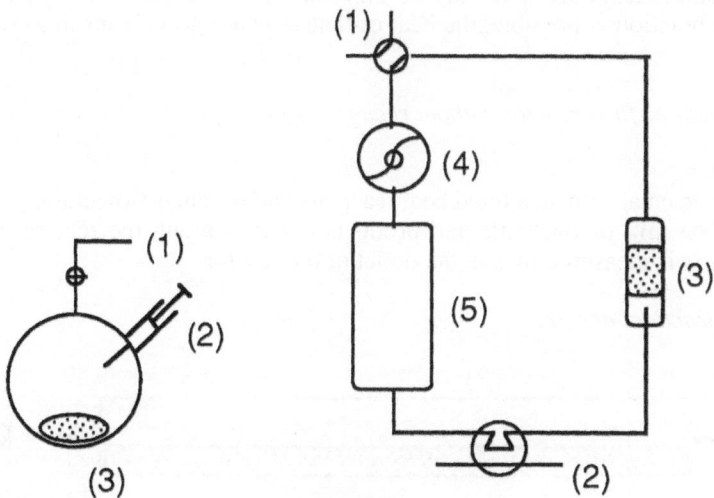

A batch   A loop reactor
(1) gas inlet, (2) Sample for analysis, (3) Catalyst
(4) recycle pump, (5) reactant volume.

c)Results.

In this reactor residence time = time on stream =t

Conversion, yields and selectivity are directly measured as function of time.

It is not so easy to obtain reaction rate (slope of the curve of conversion and yields).

It is very dangerous to draw the curve rate = f(reactant concentration) from only one run because the inhibiting effect of products is coupled to the conversion and because catalyst deactivation is not detected.

d) Advantages.

Very simple design,

Possibility to measure very small rates (long residence time),

Large conversion may be observed,

These devices may work under pressure,

The reaction scheme may be drawn from only one run,

Direct measurement of yields and conversion,

Up scaling to industrial scale is not too difficult.

e) Disadvantages.

Determination of the zero time is sometime difficult.

A good mixing of the reactants and catalyst is not easy.

Pressure or volume is not constant if the reaction changes the mole number in the gas phase.

There is no direct observation of the catalyst deactivation.

Rate determinations use slope measurements.

f) Security.

The reactant volume may be important, it is necessary to be sure that no explosive reaction is possible, the heat exchange is not good, a mean to stop racing is useful.

4.2.2. Fixed bed flow reactor without recirculation

a) Principle.

The catalyst is in a fixed bed, reactants and products flow through the bed.

The mix of reactants and products changes along the reactor. Often the composition is measured only at the outlet of the reactor.

b) Some practical design.

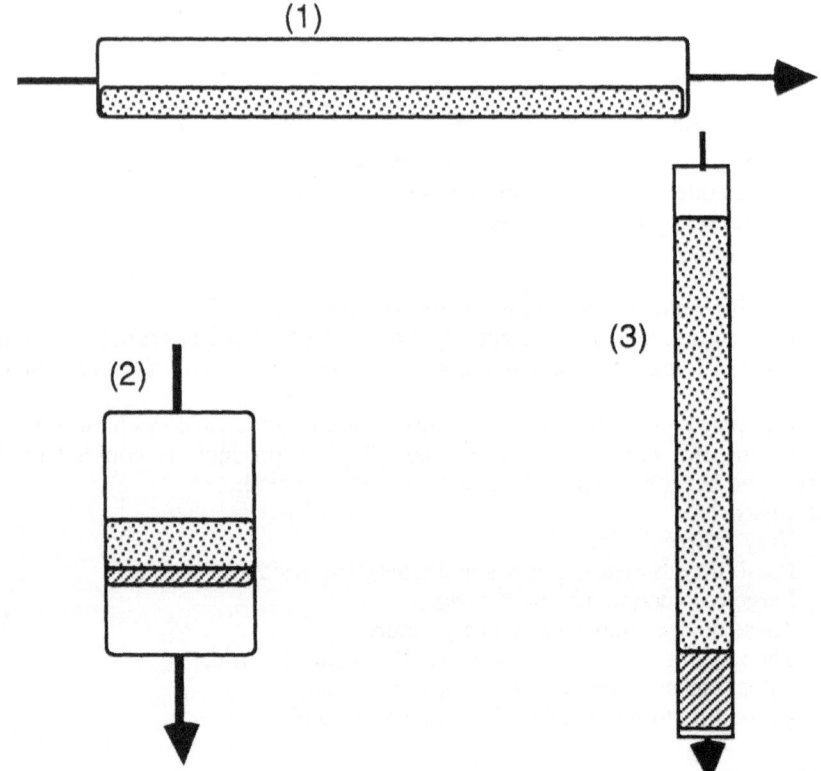

Reactor with fluid phase flowing along the bed (1) or across the bed (2, 3). Thin catalyst bed (2), Tubular reactor (perhaps plug flow reactor) (3).

c)Treating the results.

Conversion and yields are obtain from the analysis at the reactor outlet.
No local rate measurement,
No local concentration measurement,

The stability of the catalyst activity is directly observed.

d) Advantages.

The design is simple.

The tubular reactor gives a good heat exchange (for exothermic reaction).

The detection of activation or deactivation phenomena is obvious.

e) Disadvantages.

Pressure-drop may be important in the tubular reactor with powder.

There is a hot point in the reactor with exothermic reaction,

No data on the local rates of reaction is obtained. It is possible to solve this problem with a great number of sample points along the reactor.

f) Security.

The reactor may be plugged by powder, and catalyst may be blown out of the reactor.

4.2.3. Particular case: differential flow reactor

a) Principle.

It is a flow reactor with the following hypothesis:

- the conversion is small so that the reactant concentrations do not vary very much between the inlet and the outlet of the reactor, they are known,

- the product concentrations are very small and have no effect on reaction rates.

A reactor may be considered as differential when these two conditions are realised.

b) Advantage.

The reaction rate is directly measured by the difference of concentration between inlet and outlet of the reactor, with known experimental conditions (reactant flow rate, temperature, partial pressures, catalyst mass...).

c) Disadvantages.

The reaction rates are only measured with small partial pressure of products, the inhibiting effect of the products are not directly observed and the successive reactions are not studied. They may be studied when products are added to reactants.

When the reaction is exothermic and the catalyst bed is thin, the thermal exchange may be very bad and the catalyst temperature not known..

4.2.4. Perfectly mixed flow reactor

a) Principle.

The catalyst bed is in a recirculating loop. If the recirculating flow rate D is large behind the reactant flow rate d, the composition of the reactant and products mix is almost the composition of the reactor outlet and is known by the analysis. The molar flow rate at the outlet are also obtained by the d measurement and the analysis. As the composition is constant in the reactor, it is a direct measurement of reaction rate in known conditions (the best for kinetics).

b) Practical design.

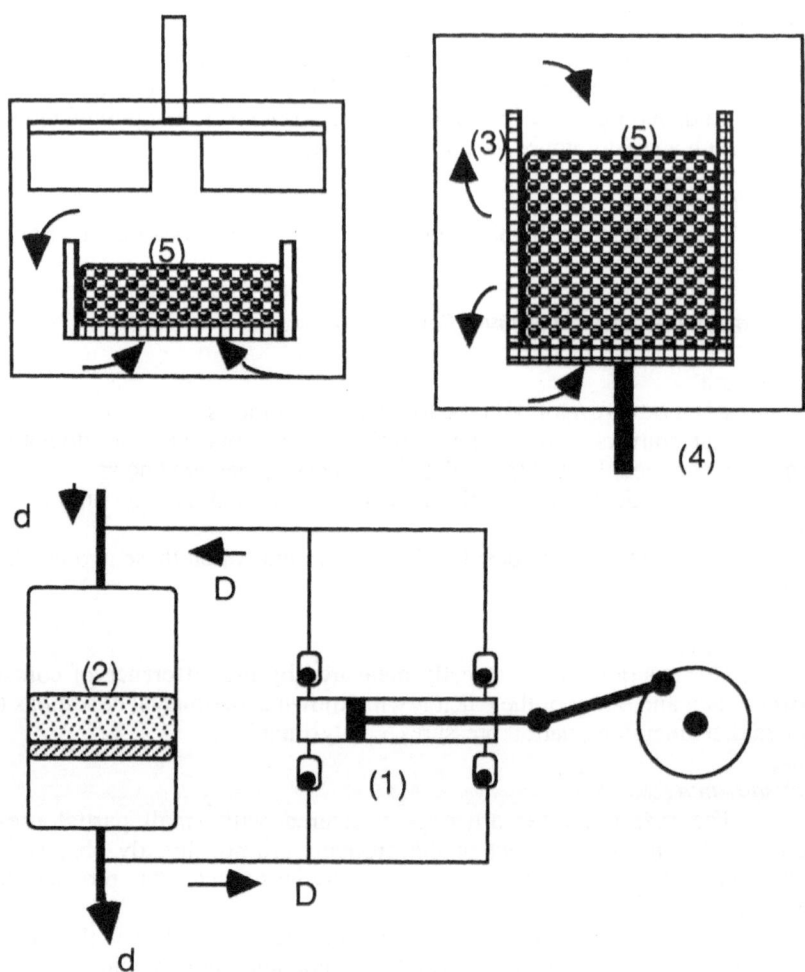

(1) circulating pump, (2) catalyst bed, (3), (4) rotating basket, (5) catalyst in pellets.

c) Advantages.

      Direct measurements of reaction rate at various conversions.
      The same pellets that in the industrial units may be used.
      Thermal exchanges are good.

d) Disadvantages.

It is necessary to realise a circulating device with D>d and verify its effectiveness.

It is necessary to verify that the various material used in the reactor have no effects on the reaction course.

The devices are fragile.

e) Security.

Reactor volume is important : explosion risk.

Large RPM numbers.

4.2.5. Pulse reactor

a) Principle.

The reactor is a flow reactor or a chromatographic column. The reactants are not introduced continuously but by pluses in the reactor. The vector gas is an inert gas or one of the reactants.

b) Reactor design.

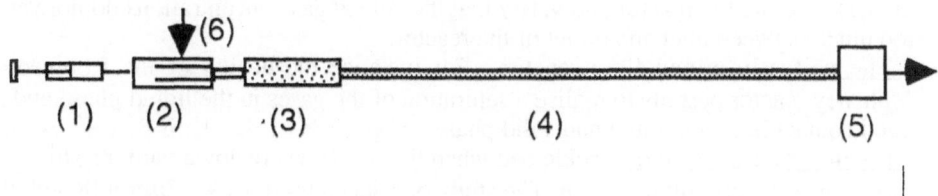

(1) reactant, (2) injector of the chromatograph, (3) catalyst, (4) chromatographic column , (5) detector, (6) vector gas

c) Advantages.

The design is very simple

Results are quickly obtained.

It may be used to observe chemisorption.

d) Disadvantages.

No reaction rate measurements (reactant concentration are not well known),

The catalyst may evolve between pulses.

e) Security.

No problem.

4.2.6. TAP (Temporal analysis of products)

This technique use very small pulses of reactants and very quick analysis by mass spectrometry.

It permits the non stationary observation of catalysis phenomena (adsorption diffusion, successive reactions ...) [6 and 7].

4.2.7. Laboratory reactors for triphasic reactions with gas-, liquids- and solid catalyst

With the three phases various states are possible:
a) In a fixed bed, with various flow rates of fluid phases various case may be observed :

    1- Continuous gas phase + liquid drops + solid
    2- Continuous gas phase + trickling liquid + solid
    3- Gas bubbles + Continuous liquid phase + solid
    4- case 2 and 3 one after the other
    5- foam + solid.

b) With a powder suspension (slurry).

    1- Continuous gas phase + suspension drops
    2- Gas bubbles + Continuous suspension phase

The various reactor types are the same as previously :

- The static reactor (slurry) : The energy of stirring must be sufficient maintain the suspension and to dissolve gases in the liquid phase. It is necessary to verify that the results do not depend on the stirring speed.

- The differential flow reactor : It is possible to realise the saturation of the liquid by the gases before the reactor and verify that the solved gas concentrations do not vary too much between inlet and outlet of the reactor.

- The perfectly mixed flow reactor : For instance the design of the Robinson-Mahoney reactor permits to realise a saturation of the gases in the liquid phase and a good contact between liquid and solid phase.

- The integral flow reactor (trickle bed when the two flow are downwards) : This reactor is very difficult to design. The study of a specialised book is useful [8 and 9].

5. REACTION SCHEME

For instance during the catalytic hydrogenation of the diethyl phthalate, the following products are observed. Are some of the following reactions negligible?

To answer this question it is possible to:

    - look at the curves TT=f(TTG)
    - study each of the individual reactions
    - add each of the supposed intermediate and observe what happens.

By these methods it is possible to measure the rates of each branch of the scheme.

The reaction scheme is an experimental result of the kinetic study. It involves only observable intermediates and simple chemical reactions and no hypothetical step.

6. KINETIC MODEL

Each branch of the reaction scheme results probably of a succession of elemental steps. The observed order in a branch for each reactant are useful to determine the step(s) limiting the rate and to propose a kinetic expression for the branch. In some case of catalyst deactivation some branches produce poisons blocked on the sites.

7. CATALYST DEACTIVATION

The catalyst activity is very often not constant with time. The catalyst duration may vary from some seconds (catalytic cracking) to several years. The deactivation is taken into account in the industrial processes. It is necessary to know the possible time on stream, to know and avoid operations killing the catalyst, and to decide to use a regeneration process.

The knowledge of deactivation is difficult to obtain at the laboratory where the time on stream are small, reactants are pure. But this knowledge is very important to build and to exploit a process..

The cost of the changing of a catalyst may be very important :
- The units is stopped during several days or weeks,
- the reactor is open,
- the old catalyst is removed,
- a new catalyst is bought,
- putting the new catalyst in the reactor is often a long and difficult operation (Pressure-drops must be equilibrated in multi-tubular reactor, the catalyst packing must be very regular in trickle bed),
- The unit is started,
- The old catalyst is regenerated or some materials are recovered.

7.1. Deactivation causes
- fouling (the catalytic site are covered by some heavy products, some pores are blocked).
- active phase modification (sintering, sublimation, dissolution in the support).
- chemical transformation (poisoning).

7.2. Model of the deactivation

7.2.1. Hypothesis of the independence of the deactivation and of the reaction
This hypothesis is generally not verified, but it was used in many papers. The reaction rate may be written

$$r_A = -\frac{dN_A}{dt} = f(T, p_A, p_B, \ldots) \cdot \Phi(t')$$

where t' is the time on stream.

The $\Phi(t')$ function is supposed independent of working conditions of the catalyst (T, partial pressure...). It is well known that these conditions have effects on the catalyst deactivation rate. Even when units work in very constant conditions the Φ function do not take into account the effects of random impurities, starting conditions of the process. Even in sintering these effects may be important on the function $\Phi(t')$

It is then supposed by analogy with kinetic order that

$$-\frac{d\Phi}{dt'} = \alpha \cdot \Phi^n \qquad \text{then} \qquad \frac{d\Phi}{\Phi^n} = -\alpha dt'$$

if $n = 1$

$$\Phi = e^{-\alpha t'} \qquad \Phi = 1 \text{ when } t'=0 \text{ et } \Phi \to 0 \text{ when } t' \to \infty$$

if $n \neq 1$

$$\frac{1}{-n+1}\Phi^{-n+1} = -\alpha t' + C \qquad \text{when } t'=0, \Phi = 1 \text{ and } C = \frac{1}{1-n}$$

$$\Phi = \frac{1}{\left(1 - (1-n)\alpha t'\right)^{1-n}}$$

The shape of the curve Φ depends of 2 parameters α, n . When these parameters are varying with reaction conditions, it is a proof that this model is not reliable.

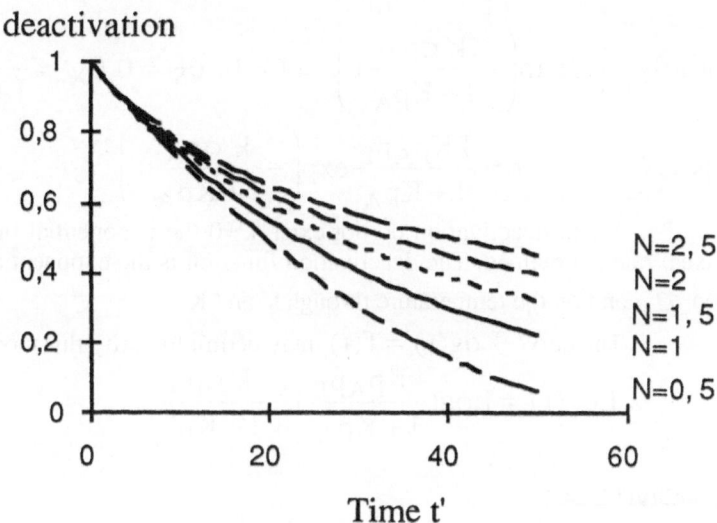

Even when searching to do an extrapolation at time zero this model is dangerous.

7.2.2. Fouling kinetic

A good method is to consider the fouling as a particular step in a general kinetics.

The fouling is the result of some steps making products strongly adsorbed on the catalytic sites, so that there are less and less active sites on the catalyst surface.

If the catalyst initially contains Nt sites and if Nc are blocked :

$$\Phi = \frac{N_t - N_c}{N_t} = 1 - \theta_c$$

Example n°1 :

The reactant contains an impurity S so that: $\dfrac{p_S}{p_A} = \sigma$

The following steps are supposed :

$$A + X \Leftrightarrow AX \qquad \theta_A = K p_A \theta_X$$
$$AX + B \rightarrow C + X \qquad r = k\theta_A p_B = kK p_A p_B \theta_X$$

$$S + X \to SX \qquad r' = k'\,\sigma p_A \theta_X = \frac{d\theta_S}{dt}$$

as $\qquad \theta_S = 1 - \theta_X - \theta_A = 1 - \theta_X(1 + Kp_A)$

$$\frac{d\theta_S}{dt} = -\frac{d\theta_X}{dt}(1 + Kp_A) = k'\,\sigma p_A \theta_X$$

and $\theta_X = C\exp\left(-\frac{k'\,\sigma p_A}{1 + Kp_A}t\right)$, à $t = 0,\ \theta_S = 0, \theta_X = \dfrac{1}{1 + Kp_A} = C$

then $\qquad r = \dfrac{kKp_A p_B}{1 + Kp_A}\exp\left(-\dfrac{k'\,\sigma p_A}{1 + Kp_A}t\right)$

When deactivation do not exist, k'=0 the exponential function =1 and the usual rate is obtained. The deactivation function is the exponential term, it depends on p_A and on the temperature through k' and K.

The curve $\text{Log}(r) = f(t)$ may permit to verify this model :

$$\text{Log}(r) = \text{Log}(\frac{kKp_A p_B}{1 + Kp_A}) - \frac{k'\,\sigma p_A}{1 + Kp_A}t$$

Example n° 2 :

Deactivation by a product of the reaction C:

$A + X \Leftrightarrow AX \qquad\qquad \theta_A = Kp_A \theta_X$

$AX + B \to C + X \qquad r = k\theta_A p_B = kKp_A p_B \theta_X$

$C + X \to CX$

$$\frac{d\theta_C}{dt} = k'\,\theta_X p_C = k'\,\theta_X \alpha v = Kkk'\,\theta_X^2 \alpha p_A p_B$$

as $\qquad \theta_C = 1 - \theta_X - \theta_A = 1 - \theta_X(1 + Kp_A)$

$$\frac{d\theta_C}{dt} = -\frac{d\theta_X}{dt}(1 + Kp_A) = Kkk'\,\alpha p_B p_A \theta_X^2$$

then $\qquad \dfrac{d\theta_X}{\theta_X^2} = -\dfrac{Kkk'\,\alpha p_B p_A}{1 + Kp_A}dt$

and $\dfrac{1}{\theta_X} = \dfrac{Kkk'\,\alpha p_A p_B}{1 + Kp_A}t + C$, when $t = 0,\ \theta_C = 0, \theta_X = \dfrac{1}{1 + Kp_A}$

then $\qquad r = Kkp_A p_B\,\dfrac{1}{\dfrac{Kkk'\,\alpha p_A p_B}{1 + Kp_A}t + 1 + Kp_A}$

The curve $\dfrac{1}{v} = f(t)$ may be useful to verify the model

$$\dfrac{1}{r} = \dfrac{k'\alpha}{1 + Kp_A} t + \dfrac{1 + Kp_A}{Kkp_A p_B}$$

Example n° 3 :

The catalytic cracking involve acidic sites X

$$A + X \Leftrightarrow AX \qquad\qquad \theta_A = Kp_A \theta_X$$
$$AX \rightarrow BX \qquad\qquad v = k\theta_A = kKp_A \theta_X$$
$$BX \rightarrow B + X \qquad\qquad \theta_B = Kp_B \theta_X$$

from $\theta_A = 1 - \theta_B - \theta_X, \quad K_A p_A \theta_X = 1 - K_B p_B \theta_X - \theta_X$

it is drawn $\theta_X = \dfrac{1}{1 + K_A p_A + K_B P_B}$

and the rate is $r = \dfrac{kK_A p_A}{1 + K_A p_A + K_B P_B}$

If θ_C sites are occupied by heavy aromatic molecules (coke) θ_A becomes

$$\theta_A = 1 - \theta_B - \theta_X - \theta_C \approx (1 - \theta_B - \theta_X)(1 - \theta_C)$$

if $\Phi(C) = 1 - \theta_C$

$$r \approx \dfrac{kK_A p_A}{1 + K_A p_A + K_B P_B} \Phi(C)$$

C is the result of some aromatisation and condensation reactions it is possible to deduce it from a kinetic expression. It is then sufficient to find an empirical expression giving $\Phi = f(C)$, for instance :

$$\Phi = 1 - \alpha C \quad \text{or} \quad \Phi = \exp(-\alpha C) \quad \text{or} \quad \Phi = \dfrac{1}{1 + \alpha C} \quad \text{or} \quad \Phi = \dfrac{b + 1}{b + \exp(aC)}$$

These expression may be determined by experiments with "coked" catalyst with a known coke contain C.

7.3. Do not miss the deactivation

The deactivation appears as an evolution with the time on stream. It may be hidden to the observer if reaction conditions are varied regularly with the time. For instance, two methods are used to change the reactor temperature to obtain the Arrhénius plot :

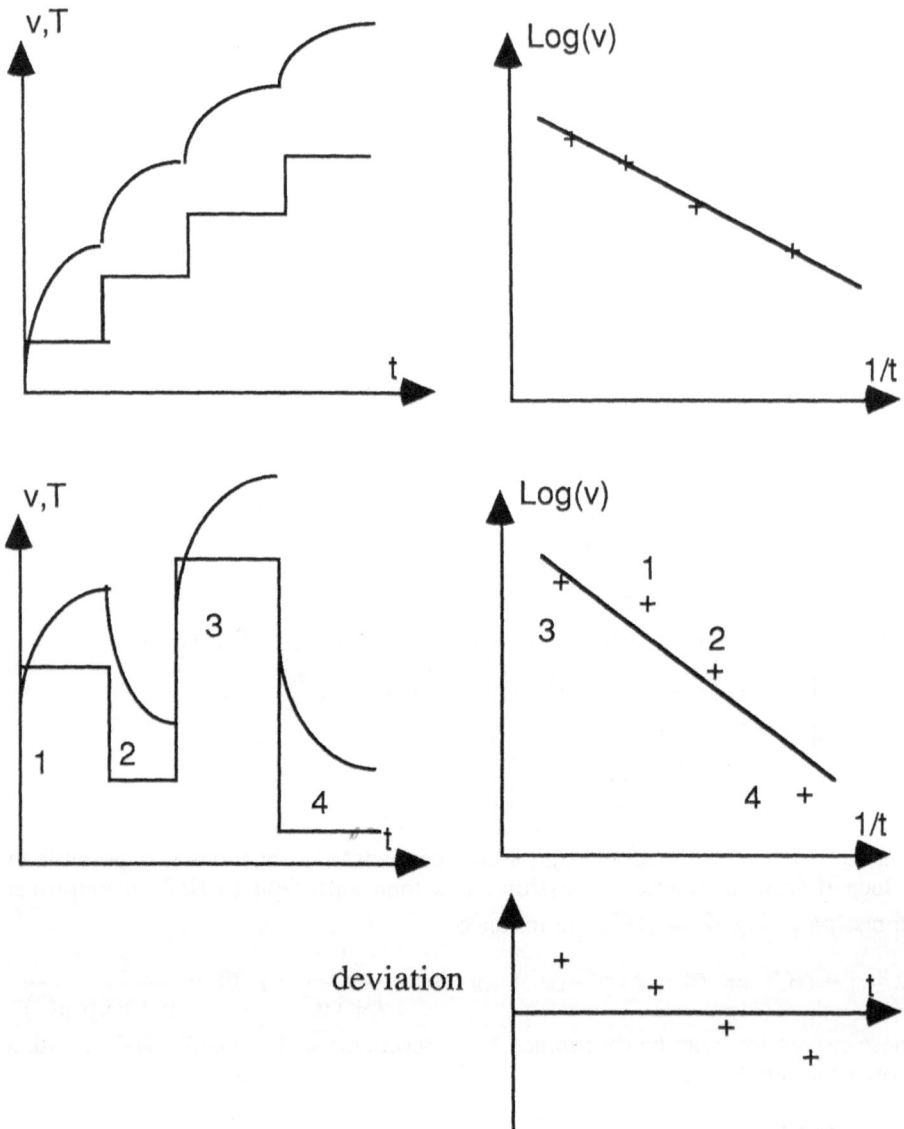

With the first method, the apparent activation energy is false, moreover nothing suggests that there is a deactivation. With the same number of experiments, the second one gives a best estimation of the apparent activation energy and the study of the deviations between the points and the best straight line show the deactivation.

8. DETERMINATION OF THE KINETIC CONSTANTS

8.1. Linearisation method

This old method is possible when the number of unknown constants is small (1 or 2). The rate expression is transformed in a linear expression

$$y = ax + b$$

where y and x are mathematical expressions containing the measurements or the known experimental conditions and no unknown constant, a an b are expressions containing the unknown constants and eventually known experimental conditions.

The quality of the model may be estimated from the alignment of the points.

When using this method it is necessary to draw the incertitude fields on the linear plot because they may vary in an unexpected field.

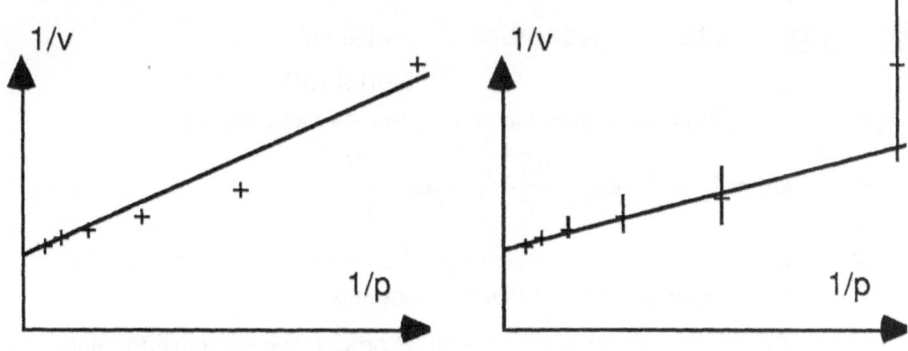

8.2. Non linear adjustment techniques

After the experiments we obtain a set of experimental values (reaction rates, yields, ...) measured in different but known conditions. The results permits also to built a model which permits to compute the same values knowing the experimental conditions and a set of unknown constants K_i (kinetic constant of steps, adsorption equilibrium constant, ...).

The problem is to find the best set of K_i constants that minimise the differences between each experimental value and the calculated one; this deviation may be characterised by a criterion.

The determination of the set of K_i is possible by an optimisation program. It contains a strategy to minimise the criterion. Some modern computer programs like "excel" contain such optimisor easy to use for simple models. For more complex problems special program may be used.

9. INTERPRETATIONS OF KINETIC CONSTANT VALUES

If the kinetic model previously determined has a physical meaning, the obtained constant values (rate constant, adsorption constants, and their activation energy) are a very good manner to characterise the catalyst, compare it to various catalysts of the same family. But researchers on catalysis often prefer to give data on site density and working frequency of sites, which may be obtained by other techniques.

Some authors like Russell W. MAATMAN, Dept. Of Chemistry, Dordt College, Sioux City, Iowa 51250, USA [11 to 13], have shown that it is possible to draw the density and the working frequency of the site from kinetic constant and activation energy.

Using the transition state theory [14] :

$$B_g \; + \; A_s \underset{k_{-1}}{\overset{k_1}{\rightleftharpoons}} BA_s \overset{K^{\#}}{\rightleftharpoons} M^{\#} \overset{k^{\#}}{\longrightarrow} \text{Produits}$$

$$\searrow \; k_3 \; \nearrow$$

gaz   site   adsorbat   état de transition

it is possible to write the expression of the observed rate constant :

$$k_3 = L \cdot \frac{kT}{h} \cdot \exp\left(\frac{\Delta S^{\#}}{R}\right) \cdot \exp\left(-\frac{E}{RT}\right) \tag{1}$$

with : L : active site density (sites by m^2 of catalyst)

k, h : Boltzmann's and Planck's constants.

$\Delta S^{\#}$: entropy variation between adsorption and the transition state.

k_3 : rate constant of the product formation from the adsorbed species, practically the experimental rate constant if adsorption is large (order 0) (en moles^{-1} m^2 of catalyst).

E : Activation energy of the step, experimentally determined

The mean working frequency of the site will be :

$$v = \frac{k_3}{L} = \frac{kT}{h} \cdot \exp\left(\frac{\Delta S^{\#}}{R}\right) \cdot \exp\left(-\frac{E}{RT}\right) \tag{2}$$

The hypothesis $\Delta S^{\#} = 0$, is not true in general.

The relation (1) bind the two measurable quantities k_3, E to two others characterising the catalyst : $L, \Delta S^{\#}$. but do not permits to draw L and $\Delta S^{\#}$. without other hypothesis.

If it is possible to obtain L by an other method (dosage, metallic area, ...) $\Delta S^{\#}$ may be determined and then the possible structure of the transition complex and perhaps the geometry of the site.

If the values of $\Delta S^{\#}$ may be estimated from an other method (Russel W. MAATMAN, gave $\Delta S^{\#}$ for 118 published reactions [13] it will be possible to estimate L.

If the site density is very small (and working frequency large) it is possible to think that active sites may be some very particular surface defects. It will be researched to increase their numbers by adding some impurities or by the preparation technique. On the contrary if the working frequency is very small (and site density large), it will be researched to increase the site activity using some electronic effects or geometric effects

When changing the reactant molecule by a molecule of the same family, or an H atom by a D atom, only $\Delta S^{\#}$ would be modified.

This kind of deduction may be amusing but do not forgot that:

- the accuracy of the experimental determinations of k_3 and E, is very often very bad,
- all the sites are somewhat different, there is a site distribution so that compensation effects will appear between the $\Delta S^{\#}$ and L values.

10. DETAILED MECHANISM AND SITE SIMULATION

The definition of the reaction mechanism is not very precise, for someone the reaction scheme or the list of elementary steps constitute a reaction mechanism. We choose a more strict definition.

The mechanism of an elementary step must show the various events that append on the catalytic site and around. As the observation of individual atoms is not possible, the mechanism is a set of hypothesis (electron transfer, hydride transfer, orbital modification, structure of adsorbed intermediates, mobility of species on the surface from a site to an other...). This description may permit to verify if the mechanism is chemically possible and if there is no contradiction between the hypothesis.

When a possible mechanism is proposed, it is useful to prospect what may happen to other molecules of the same family, what may modify the sites and the reaction course. It may suggest some experimental verifications of the proposed mechanism:

- reactivity of compound of the same family (or with some D atoms)
- effect of catalyst modification (energetic levels or geometric factors)

When the interaction between a site and a reactant is well characterised, it may be studied by theoretical computation.

11. CONCLUSIONS

The kinetic study of a reaction on a new catalyst is a long but exciting adventure. Experimental devices are not very expensive and if the work is well done, a very important knowledge of the catalyst may be obtained : reaction scheme, kinetic model, kinetic constants, catalyst site properties and reaction mechanism.

References

[1] J.E. Germain, "Catalyse de contact", Techniques de l'ingénieur, Paris, J 1181-12/1181-15 (1979).

[2] P. Trambouze, H. Van Landghem, J.P. Wauquier, Les réacteurs chimiques, conception, calcul, mise en oeuvre, Technip, Paris, 1984.

[3] J. Villermaux, Génie de la réaction chimique, conception et fonctionnement des réacteurs. Technique et Documentation, Lavoisier, Paris, 1982.

[4] C.N. Satterfield. Mass transfer in heterogeneous catalysis, MIT press 1970.

[5] Michel Boudart, G. Djega-Mariadassou, Cinétique des réactions en catalyse hétérogène, Masson, Paris, 1982.

[6] J. T. Gleaves, J. R. Ebner and T. C. Kuechler, *Catal. Rev. -Sci. Eng.*, 30 (1), 49-116 (1988).

[7] D. S. Lafyatis, G. Creten, G. F. Froment, *Applied Catal. A General*, 120 (1994)85-103.

[8] P. A. Ramachandran and R. V. Chaudhari, Three phase catalytic reactor, Gordon and Breach Science Publishers, Philadelphia, Pennsylvania, USA 1983 and 1992).

[9] A. Gianetto and P. L. Silveston, Multiphase chemical reactors, theory, design, scale up. Hemisphere Publishing Corporation, 1986.

[10] G. F. Froment, Proc 6th International Congress on the Catalyse, G.C. Bond, P. B. Wells, F. C. Tomkins ed., The Chemical Sciety, London 1977, p10-31.

[11] R. W. Maatman, *Catal. Rev.- Sci. Eng.*, 8, 1, (1973).

[12] R. W. Maatman, Advances in Catalysis **Vol. 29** p. 97. Academic Press, New York, 1980.

[13] R. W. Maatman, Site densities in unimolecular, solid-catalysed reactions, *J. Catal*, 72, 31-36, (1981).

[14] S. Gladstone, K. Laidler, H. Eyring, The theory of rate processes, Mc Grow-Hill, London, 1941.

LECTURE 4

Electronic Structure of Metals and Alloys: from Bulk to Surfaces and Clusters

G. Tréglia

C.R.M.C.²-C.N.R.S., Campus de Luminy, Case 913,
13288 Marseille cedex 9, France

---

1. INTRODUCTION

The peculiar reactivity of a catalyst depends on the *local* electronic structure that the chemisorbed molecule of interest will find at its surface: electronic structure, but also atomic and (in the case of alloys) chemical ones. It is then essential to understand how these structures are coupled to one another and how they vary with the metal species, the concentration, the orientation of the surface for semi-infinite materials and/or the cluster size for finite ones. Modelling these phenomena would indeed allow us to design the best suited catalyst for a given reaction by adjusting some of the above mentioned parameters.

The aim of this lecture is to give the tools for characterizing the electronic structure of metallic catalysts, and to show how they can be used to predict both their atomic and chemical structures. These electronic structure methods extend from *ab initio* calculations to *semi-phenomenological* models such as *pseudopotential* theory for normal metals or *tight-binding* approximation for transition metals. Since the most commonly used catalysts are metals of the end of the transition series, we will put some emphasis on the latter by giving some details on moment and continued fraction methods. The examples will also been chosen in that way, with some partiality for palladium (when possible) which can be considered as the metallic catalyst archetype. We will underline how the electronic structure is modified at surface and cluster sites, first for pure metals and then for bimetallic systems (*CPA: Coherent Potential Approximation*). Then, we will show how the energetics of the system (cohesive energy, surface tension, mixing energy) can be derived from electronic structure by using more or less sophisticated many-body potentials (*SMA: Second Moment Approximation, TBIM: Tight-Binding Ising Model*). This will allow us to get trends as a function of the number of valence electrons for various properties such as the crystalline structure of pure metals, the relaxation or reconstruction of surfaces, the shape of clusters and finally the chemical structure of bulk (tendency to ordering or phase separation) and finite (surface or site segregation) systems. In turn, we will illustrate the dependence of the local densities of states with respect to the equilibrium (geometrical, chemical) environment defined as above.

2. BULK ELECTRONIC STRUCTURE

2.1 Pure metals

2.1.1 One electron approximation (ab initio methods)

<u>2.1.1.1 Hartree(-Fock) approximation</u>: due to the mass difference between the electrons and protons, we can decouple their respective movements (*adiabatic approximation*). The main problem is then the treatment of the Coulomb interaction between the electrons, which is a quantic many-body problem involving at least all the valence electrons (external shells). The approximate way to solve this problem is to use a mean-field approximation in which a given electron interacts with all the others by means of an effective mean field $V^{eff}(r)$:

$$V^{eff}(\vec{r}) = V^{ion}(\vec{r}) + V^{H}(\vec{r}) \tag{1}$$

The first term $V^{ion}(r)$ is the ionic potential and $V^{H}(r)$ the Hartree potential:

$$V^{H}(\vec{r}) = \int d\vec{r}\,' \frac{n(\vec{r}\,')}{|\vec{r} - \vec{r}\,'|} \tag{2}$$

where n(r) is the electronic density of states, taking the electron charge lel equal to unity. The problem then reduces to a "one electron" hamiltonian:

$$H = E^{kinetic} + V^{eff} = -\frac{\hbar^2}{2m}\nabla^2 + V^{eff} \tag{3}$$

the eigenfunctions and eigenvalues of which are solutions of the Schrödinger equation:

$$H\psi_\alpha = \varepsilon_\alpha \psi_\alpha \tag{4}$$

The ground state of the system at T=0 K is then obtained by stacking the electrons in the lowest energy states available, leading to N-body states characterized by the occupation numbers (or Fermi functions) f_α which are such as $f_\alpha=1$ if $\varepsilon_\alpha < E_F$ (E_F=Fermi level) and $f_\alpha = 0$ otherwise. The spatial density of states is defined as:

$$n(\vec{r}) = \sum_\alpha f_\alpha |\psi_\alpha(\vec{r})|^2 \tag{5}$$

whereas the electronic density of states n(E), which counts the number of states with energy E, is defined as:

$$n(E) = \sum_\alpha \delta(E - \varepsilon_\alpha) = Tr\delta(E - H) \tag{6}$$

where the operator $\delta(E-H)$ is defined by: $\delta(E-H)|\psi_\alpha\rangle = \delta(E-\varepsilon_\alpha)|\psi_\alpha\rangle$ and the trace (Tr) is performed on the electronic states α. Using the mathematical properties of δ-functions, and defining the Green function G as:

$$G(z) = \frac{1}{z-H} \qquad (7)$$

(6) is rewritten:

$$n(E) = \lim_{\eta \to 0^+} \left[-\frac{Im}{\pi} Tr\{G(E+i\eta)\} \right] \qquad (8)$$

Solving the Schrödinger equation (4) is then performed self-consistently by starting with trial density $n(r)$ which allows to calculate $V^H(r)$, then new ψ_α and ε_α and then new $n(r)$ ... up to convergency. Let us just recall two common features to all mean-field approximations:
- the total band energy of the sytem is not equal to the sum over the one electron energies (ε_α), since it counts twice the electron-electron interactions, which then have to be subtracted once

$$E_0^H = \sum_\alpha f_\alpha \varepsilon_\alpha - \frac{1}{2} \int d\vec{r} V^H(\vec{r}) n(\vec{r}) \qquad (9)$$

- the variational properties of the energy are preserved, which means that $\delta E_0^H = 0$ to first order with respect to variations of $n(r)$.

The total energy is then obtained by adding the ion-ion contribution to the band one:

$$E_0 = E_0^H + \frac{1}{2} \int d\vec{r} V^{ion-ion}(\vec{r}) n_{ion}(\vec{r}) \qquad (10)$$

$$V^{ion-ion}(\vec{r}) = \int d\vec{r}\,' \frac{n_{ion}(\vec{r}\,')}{|\vec{r} - \vec{r}\,'|}$$

where $n_{ion}(r)$ is the ionic density: $n_{ion}(r) = Z\,\delta(r-n)$, for Z charges at sites n.

One problem of this Hartree approximation is that it does not account for the Pauli principle since it uses symmetric wave functions. It is possible to go beyond by using antisymmetric functions. This is the Hartree-Fock approximation which corrects the Hartree potential by including a so-called *exchange* contribution, which damps the Coulomb potential contribution for parallel spins. Unfortunately, this is a rather dissymmetric way to treat the electron interactions between the electrons since all the electrons should avoid one another. Therefore, whereas the electronic correlations are completely neglected in Hartree, they are treated in a too much dissymmetric way in Hartree-Fock.

2.1.1.2 Local Density Functional Approximation (LDA): it is then tempting to treat the correlations in a more symmetric way, which can be done in the framework of the Local Density Functional Approximation [1], which up to now is the most widely used *ab initio* method. It is grounded on the assumption that the ground state energy E_0 can be written as a functional of the density of states n(r), $E_0=E_0[n(r)]$, which is minimum for the real density of the system. This leads to write n(r) under the same "one electron" form as (5), using wave functions ψ_α which are solutions of an Hamiltonian similar to (3), but with an effective potential:

$$V^{eff}(\vec{r}) = V^{ion}(\vec{r}) + V^H(\vec{r}) + V^{xc}(\vec{r}) \tag{11}$$

which differs from (1) by the introduction of an *exchange-correlation* term V^{xc}, which is the functional derivative of a contribution $E^{xc}[n(r)]$ to $E_0[n(r)]$:

$$V^{xc}(\vec{r}) = \frac{\partial E^{xc}[n(\vec{r})]}{\partial n(\vec{r})} \tag{12}$$

All the difficulties are then transfered in this term which has to be approximated. The usual way is to assume that E^{xc} is a local functional of n(r), i.e. that it is defined from the knowledge of the density at $\vec{r}$ only. That means that the first corrections to (7) should involve the gradient of the density: $\vec{\nabla}n(\vec{r})$.

$$E^{xc}[n(\vec{r})] \approx \int n(\vec{r})\varepsilon^{xc}[n(\vec{r})]d\vec{r} \tag{13}$$

$\varepsilon^{xc}(n)$ is the so-called "exchange and correlation" energy of a uniform electron system with density n. The total energy then writes as:

$$E_0^{LDA} = E_0 + \int d\vec{r}n(\vec{r})\left(\varepsilon^{xc}[n(\vec{r})] - \frac{\partial\varepsilon^{xc}[n(\vec{r})]}{\partial n(\vec{r})} \right) \tag{14}$$

where E_0 is taken from (10). This requires to know $\varepsilon^{xc}(n)$ which is in general taken as the average value of the exchange potential for a free electron system ($\propto n^{1/3}$), weighted by an empirical factor α (X_α method). In practice, one uses approximate one electron potentials, among which the most commonly used is the "muffin tin" one, for which the potential is calculated exactly in spherical regions centered around the nucleus whereas it is taken equal to zero in the intersticial region. The corresponding orbitals are then linearized leading to the LMTO (Linear Muffin Tin Orbital) approximation [2], from which some results will be used in the following.

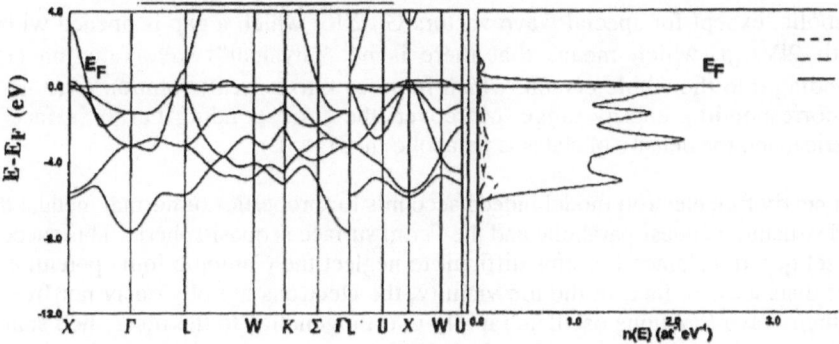

Figure 1: LMTO calculation of the band structure and density of states of Pd [3].

The LDA method gives very good results concerning both the band structure and density of states (see figure 1 for Pd) and for the lattice parameter, elastic constants and cohesive energies (at least when gradient corrections are taken into account). This is very satisfying since this method is *ab initio*, i.e. "without parameters" (contrary to more empirical methods which will be developed later), which does not mean that it is "without approximation" as shown above ! Nevertheless, this method remains less suited for non periodic systems, in presence of defects, and tedious to use coupled with numerical simulations such as Molecular Dynamics ... even though *Car-Parinello* type methods [4] are now developed which take into account simultaneously the movements of ions and electrons. But such methods remain heavy to handle for large systems, which justify to develop simpler methods, using semi-empirical potentials suited to the system under study.

2.1.2 Normal metals (jellium, pseudopotentials)

For the metals of the columns I, II, III of the periodic classification, the bonding involves conduction electrons of s or p type which have a strongly delocalised character. They can then be treated as (nearly) free electrons, moving without mutual interactions in a jellium of positive charges, the lattice potential being then introduced as a perturbation. The eigenfunctions of the Schrödinger equation (3-4) are then Bloch waves, i.e. plane waves modulated by a function $u_k(\vec{r})$ having the lattice periodicity:

$$\phi_k(r) \propto exp\, i\vec{k}\vec{r} \;\; u_k(\vec{r}) \tag{15}$$

with eigenenergies:

$$\varepsilon_k = \frac{\hbar^2 k^2}{2m} \quad (\pm \left| V_{\vec{G}_0} \right| \; if \; \vec{k} = \frac{\vec{G}_0}{2}) \tag{16}$$

where G_0 is a reciprocal lattice vector and V_{G_0} the corresponding Fourier component of the periodic lattice potential. Therefore the band dispersion E(k) is parabolic, except for special wave vectors $G_0/2$ for which a gap is opened with a width $2|V_{G_0}|$, which means that there is no "physical" wave function (i.e. extending into the whole crystal, which imposes k to be real) solution of (3-4) in the corresponding energy range. Moreover, the corresponding Fermi surface is spherical and the density of states is parabolic ($n(E) \propto \sqrt{E}$).

This nearly free electron model indeed accounts for properties of normal metals: the band structure is quasi-parabolic and the Fermi surface is quasi-spheric. This success can set questions since it seems difficult to neglect the Coulomb ionic potential at short distances. In fact, in the ion vicinity, the electrons are obviously not free ... but their wave functions oscillate rapidly to orthogonalize to the inner shell states, leading to a large kinetic energy which almost compensates the potential energy. One can then define weak **pseudopotentials** associated to pseudo-nearly free wave functions (Ashcroft empty core [5]), which indeed can be treated within the nearly free electron approximation.

2.1.3 Transition metals (tight-binding approximation)

The tight-binding method starts from the opposite point of view since it starts from isolated atoms with discrete levels, which form energy bands when the atomic wave functions overlap ... but not too much ! It assumes that any one electron electronic state $\psi(r)$, delocalised in the solid, can be written as a linear combination of atomic orbitals (LCAO): $\phi_\lambda(r-n) = |n, \lambda\rangle$ where λ labels the orbital at site n [6]:

$$\psi(r) = \sum_{n,\lambda} a_n^\lambda \phi_\lambda(r-n) \tag{17}$$

which is the more justified as the overlap between the orbitals is weak (d states of transition metals, sp valence electrons of semi-conductors, ...). This means that the potentiel V^{eff} in the Hamiltonian is replaced by the sum of the atomic potentials at sites n: V_n^{at}. Then for a non degenerate ("s" type) state, the Hamiltonian writes:

$$H = \sum_n |n\rangle(\varepsilon^{at} + \alpha)\langle n| + \sum_{n,m} |n\rangle\beta(m-n)\langle m| \tag{18}$$

with matrix elements: $<n|H|n> = \varepsilon^{at} + \alpha$ and $<n|H|m> = \beta(m-n)$

- ε^{at} : atomic level $\varepsilon^{at}|\phi_i\rangle = \left(-\dfrac{\hbar^2}{2m}\nabla^2 + V_i^{at}\right)|\phi_i\rangle$

- α : crystalline field integral $\alpha = \left\langle\phi_i\left|\sum_{n\neq i} V_n^{at}\right|\phi_i\right\rangle$

- $\beta(m-n)$: hopping integral $\beta(m-n) = \left\langle\phi_n\left|V_n^{at}\right|\phi_m\right\rangle$

The hopping integral allows the electrons to jump from site to site, and lifts the degeneracy of the atomic states. It only depends on the distance (m-n) between sites n and m: $\beta(m-n) \sim e^{-q(m-n)}$, and is rapidly damped (after 1st or 2nd neighbours). If one reintroduces the degeneracy of d orbitals, one has to define a matrix with elements: $\beta_{nm}^{\lambda\mu}$. Due to the spherical symmetry of atomic potentials, this matrix is diagonal in the basis of spherical harmonics with the z-axis along (m-n), with eigenvalues defined as the integrals σ, π, δ, ... following the quantum magnetic number $|m|=0, 1, 2, ...$ This leads to different hopping integrals labelled $ss\sigma$, $pp\sigma$, $pp\pi$, $dd\sigma$, $dd\pi$, $dd\delta$, ... For the d-d interactions, one can define canonical parameters such as: $dd\sigma<0$, $dd\pi>0$ and $dd\delta<0$ (since $V^{at} < 0$) and $|dd\sigma| \approx 2 |dd\pi|$ et $dd\delta \approx 0$ [7].

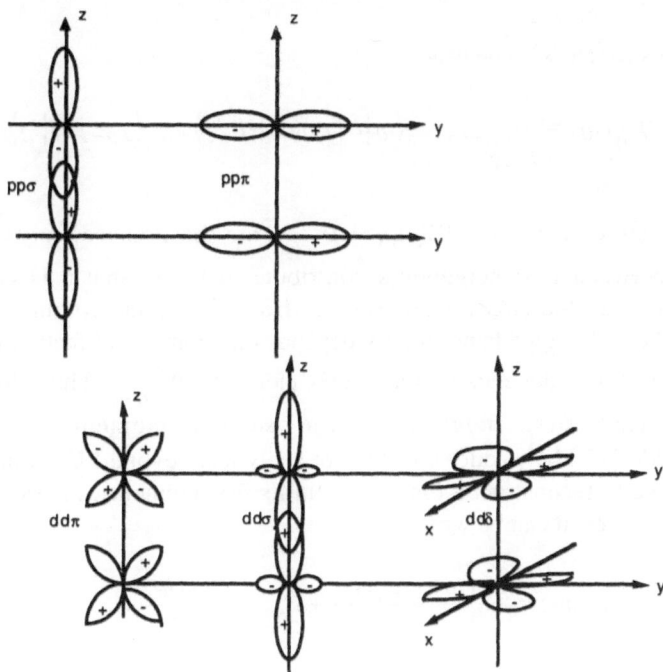

Figure 2: Schematic p-p and d-d orbitals.

For a non degenerate "s" band, the Bloch theorem is sufficient to find the energy levels ε_k, whereas for a degenerate d band, one has to diagonalize a (5x5) matrix. This gives $\varepsilon(k)$ for wave vectors k parallel to high symmetry directions in the Brillouin zone. Then from the dispersion curve $\varepsilon(k)$ one gets the density of states (which is sufficient to calculate band energy). But the essential advantage of the method is that it allows to work in the direct space to calculate densities and energies, without resorting to diagonalisation of the hamiltonian, and then without need for the Bloch theorem, which allows to deal with non crystalline solids and defects. Indeed, one can derive n(E) from Tr δ(E-H) which can be calculated over any basis, and in particular in the basis of atomic orbitals |n>. More precisely, one can introduce the local density of states at site i_0:

$$n_{i_0}(E) = -\frac{Im}{\pi}\left\langle i_0\left|G(E+i0^+)\right|i_0\right\rangle \tag{19}$$

Since the Green operator is difficult to calculate, let us define the moments of this local density of states:

$$\mu_p(i_0) = \int\limits_{-\infty}^{+\infty} dE E^p n_{i_0}(E) = \left\langle i_0\left|H^p\right|i_0\right\rangle \tag{20}$$

which writes for an "s"-type band:

$$\mu_p(i_0) = \sum_{j_1,j_2,\ldots,j_{p-1}} \left\langle i_0|H|j_1\right\rangle\left\langle j_1|H|j_2\right\rangle\ldots\left\langle j_{p-1}|H|i_0\right\rangle$$

Since the matrix elements $<j_p|H|j_{p+1}> = \beta_{j_p j_{p+1}}$ are short ranged, only hoppings on a site or between first neighbours contribute to μ_p so that it is given by the summation of all the closed paths starting from site i_0 and coming back with p hoppings. For a "s" type band, it only depends on geometrical factors whereas for the "d" one, it also depends on the matrix elements $\beta_{j_p j_{p+1}}^{\lambda\mu}$. Thus, for a site i_0 having $Z(i_0)$ first neighbours, the second moment is given by: $\mu_2 = Z(i_0)(dd\sigma^2+2dd\pi^2+2dd\delta^2)/5$. It is not analytical beyond ! Unfortunately, it is not so simple to reconstruct a function from its first moments. Indeed, developing $(z-H)^{-1}$ as a series, it can be written:

$$\left\langle i_0|G(z)|i_0\right\rangle = \frac{\mu_0(i_0)}{z} + \frac{\mu_1(i_0)}{z^2} + \ldots\ldots + \frac{\mu_p(i_0)}{z^{p+1}} + \ldots\ldots \tag{21}$$

which does not converge ! The best solution is to write the density of states as a continued fraction, which is possible if one can derive from the atomic orbital basis |n> a new basis |n} in which the hamiltonian matrix is tridiagonal, starting from the atomic orbital on which the local density is calculated $|1\} = |i_0>$. In this new basis, the tight-binding hamiltonian writes:

$$H = \sum_n |n\}a_n\{n| + \sum_n |n\}b_n\{n+1| + \sum_n |n\}b_{n-1}\{n-1| \tag{22}$$

$$\{n|H|n\} = a_n; \quad \{n|H|n+1\} = b_n; \quad \{n|H|n-1\} = b_{n-1}; \quad \{n|H|n \pm p\} = 0$$

so that the Green function $G(z) = (z-H)^{-1}$ is obtained as:

$$G(z) = \frac{1}{D_0} \begin{bmatrix} D_1 & \cdots & \cdots \\ \cdots & \cdots & \cdots \\ \cdots & \cdots & \cdots \end{bmatrix} \quad \text{with} \quad D_i = \begin{vmatrix} z - a_{i+1} & -b_{i+1} & 0 & \cdots \\ -b_{i+1} & z - a_{i+1} & -b_{i+2} & \cdots \\ 0 & -b_{i+2} & z - a_{i+1} & \cdots \\ & \cdots & & \cdots \end{vmatrix}$$

which is expanded using the recursion law: $D_i = (z - a_{i+1})D_{i+1} - b_{i+1}^2 D_{i+2}$. It is then easy to project G on the first orbital: $G_{11} = \{1|(z-H)^{-1}|1\} = <i_0|(z-H)^{-1}|i_0>$, which leads to a continued fraction [8]:

$$G_{11} = \frac{D_1}{D_0} = \cfrac{1}{z - a_1 - \cfrac{b_1^2}{z - a_2 - \cfrac{b_2^2}{z - a_3 - \cfrac{b_3^2}{\cdots\cdots}}}} \tag{23}$$

One can show by identifying (21) and (23) that the 2N 1st moments of the density of states are related to the N first couples of coefficients $(a_{n<N}, b_{n<N})$: $\mu_0 = 1$; $\mu_1 = a_1$; $\mu_2 = a_1^2 + b_1^2$; $\mu_3 = a_1^3 + 2a_1 b_1^2 + a_2 b_1^2$; .... . As a consequence, one could derive the continued fraction coefficients from the moments obtained by counting paths on the lattice, but it is numerically ill conditionned. It is better to calculate directly the coefficients by constructing the new basis $|n\}$ tridiagonalising H, which is easy using a so-called *recursion* method [9]:

$$\|1\} = |1\rangle$$
$$\|2\} = H\|1\} - a_1\|1\}$$
$$\cdots\cdots\cdots\cdots\cdots\cdots\cdots$$
$$\|n+1\} = H\|n\} - a_n\|n\} - b_{n-1}^2\|n-1\}$$
$$\tag{24}$$

The orthogonalisation relations then give:

$$a_n = \{n|H|n\} \quad ; \quad b_{n-1} = \{n-1|H|n\} \quad \text{with} \quad |n\} = \frac{\|n\}}{\sqrt{\{n\|\|n\}}} \tag{25}$$

The problem is then to terminate the continued fraction. From the definition of the new basis, the calculation of $G_{11}(E+i0)$ requires to build a cluster centered on site i_0, with nearest neighbours of i_0, then neighbours of the latter, and so on ... up to a cluster made of N shells for N pairs of coefficients. Then,

- for a finite cluster, $b_n \rightarrow 0$ beyond a given level so that the function is truncated leading to a discrete spectrum for $n_{11}(E) = -$ Im $G_{11}(E+i0)$ made of δ-functions broadened by the imaginary part η.

- for a bulk material, the coefficients converge towards values a_∞ and b_∞ which are related to band edges, at least for a band without gap [10]. In that case, the termination of the fraction reduces to the function $\Gamma(z)$, solution of the second degree equation $\Gamma(z)=(z-a_\infty-b_\infty^2\Gamma(z))$, which has an imaginary part for z=E only if: $a_\infty-2b_\infty<E<a_\infty+2b_\infty$. The density of states is then that of a finite cluster, embedded in an effective infinite medium with a semi-elliptic density of states centered in a_∞ with a width $4b_\infty$, which naturally broadens the delta functions. One can then fit the asymptotic coefficients to the band structure. Obviously, the density of states will be the more precise as the number of exact coefficients is large. This is illustrated in figure 3 in the case of Pd. It is worth noticing that in this case (end of the transition series), it is necessary to take into account the s and p valence electrons and their hybridization with the d ones to get a density of states in good agreement with that derived from ab initio calculations (fig. 1).

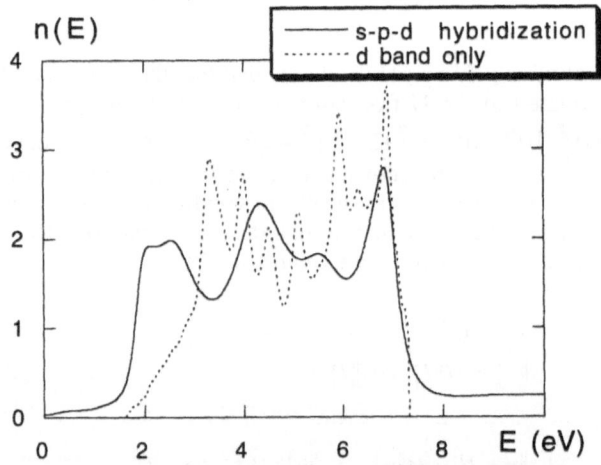

Figure 3: Pd bulk density of states, with and without sp-d hybridization, calculated in the tight-binding framework with 20 exact couples of coefficients [11].

2.1.4 Energetics and structure

2.1.4.1 Normal metals: in that case the ionic potentials are replaced by pseudopotentials, which act as perturbations on the jellium of density n_0. The new electronic density is then $n(r)=n_0+\Delta n(r)$ and the ionic one $n_{ion}(r)=-n_0+\Delta n_{ion}(r)$. The new wave functions are then written:

$$\psi_k(r) = \varphi_k(r) + \sum_{q\neq 0} \varphi_{k+q}(r)\frac{V_q}{\varepsilon_k - \varepsilon_{k+q}}$$

Expliciting the density variation, the energy (10) separates in two terms [12]:

$$E_0 = U_0 + U_{bs} \tag{26}$$

where the first term:

$$U_0(R_{at}) = \sum_{|k| \le k_F} \frac{\hbar^2 k^2}{2m} + \frac{1}{2} \int \frac{(n_{ion}(r) + n_0)(n_{ion}(r') + n_0)}{|r - r'|} \, dr \, dr'$$

is the energy of a system made of neutral spheres with radius R_{at} embedded in a uniform electron gaz with density n_0. This term is sufficient to determine the atomic radius R_{at} (at about 10%).

The second term:

$$U_{bs} = \sum_{q \ne 0} |S(q)|^2 \chi(q) V^*(q) V^{ion}(q) \approx - \sum_{|G| \le 2k_F} |S(G)|^2 |V^{ion}(G)|^2$$

$$\text{with:} \quad \chi(q) = \sum_{k \le k_F} \frac{1}{\varepsilon_k - \varepsilon_{k+q}} \qquad \text{(positive susceptibility)}$$

is the band term which couples the ionic bare potential $V_{ion}(q)$ to the screened one $V(q) = V^{el}(q) + V^{ion}(q)$. $S(q)$ is the structure factor. This is this perturbative term U_{bs} which determines the crystalline structure in order to maximize the number of reciprocal lattice vectors below $2k_F$, avoiding the q-range where $V^{ion}(q)$ almost vanishes. The crystalline structure then depends on the valence band filling via $2k_F$ $\sim N_e^{-3}$: mono and divalent metals are then found hcp, trivalent fcc, alloys between mono and divalent cc.

It can be convenient to rewrite the energy in real space under the form:

$$E_0 = \tilde{U}_0 + \frac{1}{2} \sum_{i,j} V(R_i - R_j) \tag{27}$$

$$V(|\vec{R}_i - \vec{R}_j|) = \frac{2}{N} \sum_{\vec{q}} \frac{\chi(q)}{1 + \frac{4\pi^2}{q^2} \chi(q)} (V^{ion}(q))^2 e^{i\vec{q}(\vec{R}_i - \vec{R}_j)}$$

These pair interactions V(R) show "Friedel oscillations" which are damped with the distance. It is worth noticing that such a pair development of the energy is closely related to the use of a second order perturbation method, which is only justified due to the "weakness" of the pseudopotentials. It is then limited to the case of normal

metals. The best structure is then that for which the equilibrium distance coincides
with the minimum of V(R) [12].

2.1.4.2 Transition metals: in that case, the band term in (10) can be written:

$$\sum_\alpha f_\alpha \varepsilon_\alpha = \int^{E_F} En(E)dE = \sum_n \int^{E_F} (E - \varepsilon_n)n_n(E)dE + \sum_n \varepsilon_n N_n \qquad (28)$$

where $n_n(E)$ and $N_n \left(= \int^{E_F} n_n(E)dE \right)$ are respectively the local density and the
charge at site n. Assuming charge neutrality, $N_n = Z_n$ (ionic charge), the total
energy writes [13]:

$$E - E_{at} = \sum_n \int^{E_F} (E - \varepsilon_n)n_n(E)dE + \frac{1}{2} \sum_{n \neq m} \iint drdr' \frac{Q_n(r)Q_m(r')}{|r - r'|} \qquad (29)$$

where E_{at} is the atomic level. $E-E_{at}$ is then the cohesive energy E_{coh}. The first
term in the r.h.s. member is the band energy (E_{band}) and the second one the pair
interaction (E_{rep}) between neutral atoms with charge density: $Q_n(r)=Z_n \delta(r-n)-N_n(r-n)$. Unfortunately, E_{rep} is not sufficient to account for the repulsive part of the
energy. Actually, the tight-binding approximation fails to reproduce repulsions at
short distances since it does not account for the non orthogonality of wave
functions on different sites and for the compression of sp electrons which play an
important role before that the Coulomb repulsion (1/R) becomes really efficient.
To go beyond this dificulty, the idea is to build a semi-phenomenological tight-
binding model in which the band part is well justified and the repulsive one fitted to
some physical properties.

$$E_{coh} = E - E_{at} = E_{band} + E_{rep} \qquad (30)$$

Since it comes from the electronic structure, the first term has a many-body
character whereas the second one is a phenomenological pairwise one. Note that
this solution is valid in the tight-binding limit, which means as long as: $|E_{band}|$
>> $|E_{rep}|$. The underlying assumption is that the atoms are "pseudo-atoms". with
effective atomic levels determined from local charge neutrality condition, which
saves the variational properties of the energy (1st order compensation between
errors on the self-consistent determination of local charges and atomic levels ε_n).
In the extreme tight-binding limit (R large), E_{rep} becomes negligible so that the
cohesive energy reduces to:

$$E_{coh} \approx |E_{band}| \approx -\int^{E_F} dE(E - \varepsilon^{at})n(E) \qquad (31)$$

Since the integral does not depend on details of n(E), one can use a schematic
rectangular density of states with width W. One finds a parabolic variation of E_{coh}
as a function of the d band filling N_e, $E_{coh} = N_e (10-N_e) W / 20$, which

corresponds to the successive filling of the bonding then antibonding states. This behaviour fairly compares with experimental trend (see figure 4).

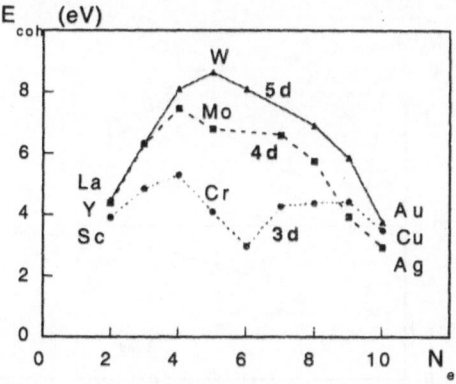

Figure 4: Experimental variation of the cohesive energies for transition metals.

Unfortunately the rectangular shape is too schematic since it brings some confusion between the band width ($n(E)\neq0$): $W\approx Zdd\beta$ and the effective width (2nd moment) which is more significant and varies as: $W_{eff}\approx dd\beta\sqrt{Z}$.

$$\mu_2 = \frac{1}{W_{eff}} \int_{\varepsilon_d-W_{eff}/2}^{\varepsilon_d+W_{eff}/2} E^2 dE = \varepsilon_d^2 + Zdd\beta^2 \qquad \Longrightarrow \qquad W_{eff} = dd\beta\sqrt{12Z}$$

This introduces the so-called second moment approximation in which $n(E)$ is fitted to the second moment of the exact density of states, leading to some universal trend for E_{band} and N_e : $E_{band}=\sqrt{\mu_2}\, f_1(E_F/\sqrt{\mu_2})$; $N_e=f_2(E_F/\sqrt{\mu_2})$ so that $E_{band}=\sqrt{\mu_2}$ $f(N_e)$. Then if one assumes that both the hopping integrals $dd\beta$ and the repulsive pair interactions exponentially decrease with distance, one finds [14]:

$$E - E_{at} = -F(N_e)dd\beta\sqrt{\sum_R e^{-2q(\frac{R}{R_0}-1)}} + A\sum_R e^{-p(\frac{R}{R_0}-1)} \qquad (32)$$

where R_0 is the first neighbour distance, $p\gg q$ and $dd\beta^2=dd\sigma^2+2dd\pi^2+2dd\delta^2$. The equilibrium lattice condition $dE/dR=0$ leads to

$$pA + qdd\beta F(N_e) = 0 \qquad (33)$$

which determines equilibrium values of R_0 and E_{coh}. This leads to the so-called second-moment interatomic potential, illustrated in figure 5 for Pd. Note that this potential is very similar to those derived elsewhere within the EAM (Embedded Atom Model [15]) or within the Glue Model [16]. Its main advantage compared to the latter is its physical transparency which clearly shows its limitations... and then its possible improvements (increasing the number of exact moments).

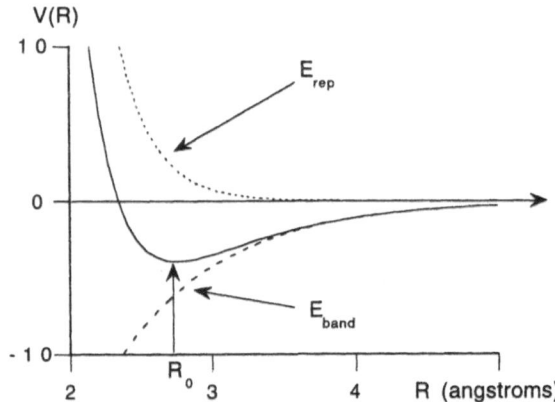

Figure 5: Distance dependence of the second moment potential for Pd.

If one assumes that q, p and ddβ slowly vary with N_e one finds that the atomic volume $V_{at}(N_e)$ is minimun for $N_e \approx 5$, the experimental asymmetry (see figure 6) being ascribed to the influence of s-d hybridization s-d. The bulk modulus varies as $E_{coh}(N_e)$, the experimental shift in the maximum (see figure 6) being due to variations of p. However some features are wrong. Actually, the bulk modulus depends significantly on the number of moments used. In the second moment approximation, it is found too weak for cc metals with Ne=4 (V, Nb et Ta) and too large for fcc ones around Ne=8.

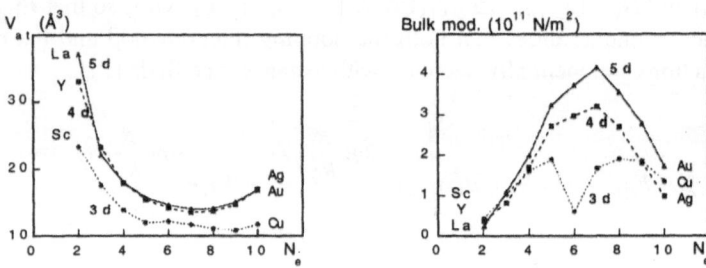

Figure 6: Experimental variation of the atomic volume and bulk modulus along transition metal series.

On the other hand, as can be seen below, the crystalline structure of transition metals is clearly related to the band filling. As shown in figure 7, the main trends are correctly reproduced by the tight-binding calculation as well for the cc/fcc as hcp/cfc systematics (which of course requires to go beyond second moment since hcp and fcc structures are identical up to second neighbours), except for the nearly filled band for which the cc structure is found instead of fcc. Fortunately, this is corrected if one takes into account sp-d hybridization which, as previously mentioned, plays a major role at the end of transition series.

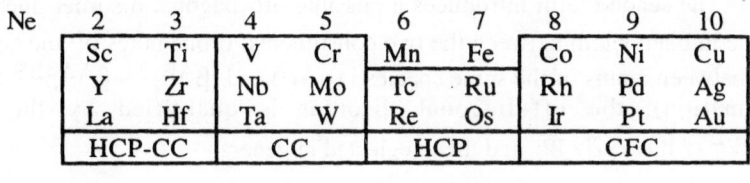

Ne	2	3	4	5	6	7	8	9	10
	Sc	Ti	V	Cr	Mn	Fe	Co	Ni	Cu
	Y	Zr	Nb	Mo	Tc	Ru	Rh	Pd	Ag
	La	Hf	Ta	W	Re	Os	Ir	Pt	Au
	HCP-CC		CC		HCP			CFC	

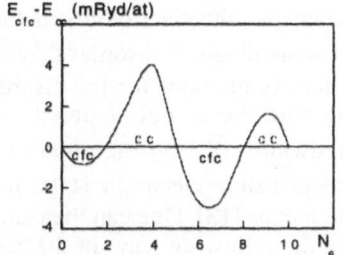 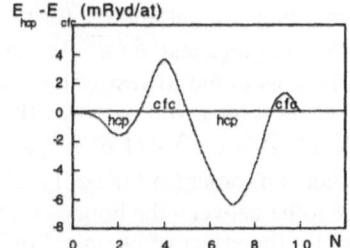

Figure 7: Stabilities of fcc relative to cc (left) and hcp (right) structures in the tight-binding framework (d band only) [13]. Note the two different orders of magnitude.

2.2 Alloy $A_c B_{1-c}$

2.2.1 Electronic structure

We will restrict ourselves to the case of transition and noble metal systems, which are the only ones used for catalysis purposes. Furthermore, for the sake of simplicity we will present the case of a "s" band (no problem to introduce degeneracy) and will assume a rigid lattice for the alloy. Therefore, one can still use the tight-binding hamiltonian (18) which has to be rewritten:

$$H = \sum_n |n\rangle \varepsilon_n \langle n| + \sum_{n,m} |n\rangle \beta_{nm} \langle m| \tag{34}$$

but it now depends on chemical configuration $\left\{ p_n^i \right\}$, $p_n^i = 1$ if site n is occupied by atom i (= A, B) and $p_n^i = 0$ if not, via:

$$\varepsilon_n = \sum_{i=A,B} p_n^i \varepsilon^i \quad \text{and} \quad \beta_{nm} = \sum_{i,j=A,B} p_n^i p_m^j \beta_{nm}^{ij} \tag{35}$$

The first term accounts for the so-called "diagonal disorder" which comes from the variation of the atomic level ε^i as a function of the element (it should also depend on environment but the smallness of charge transfers makes selfconsistency effects negligible). Along a transition series, ε^i decreases by about 1eV per element. This diagonal disorder is quantified by the ratio $\delta / \overline{W}$, where $\delta = \varepsilon^A - \varepsilon^B$ and $\overline{W}$ is the average band width ($cW^A+(1-c)W^B$). This ratio, which is somewhat related to the difference in electronegativity between the two constituents, is important since $\delta/\overline{W} \sim 1$. The second term introduces a possible off-diagonal disorder due to the difference in band width between the two constituents. If one notes β^{ii} the hopping integral between atoms of the same species i ($\sim W^i$) and $(\beta^{AB})^2 \sim \beta^{AA}\beta^{BB}$ (Shiba approximation), the off-diagonal disorder is quantified by the ratio $(W^A - W^B)/\overline{W} << \delta/\overline{W}$, and then neglected in general.

If one uses canonical parameters for the pure elements, the electronic structure of the alloy depends on c and $\delta/\overline{W}$. In the case of weak diagonal disorder ($\delta/\overline{W} << 1$), perturbations at the lowest order lead to a density of states for the disordered alloy which is almost the same as that of the pure metal, but centered on the average level: $\overline{\varepsilon} = c\,\varepsilon^A + (1-c)\,\varepsilon^B$ with a bandwidth $\overline{W}$. On the contrary, for strong diagonal disorder ($\delta/\overline{W} >> 1$), a gap is opened since electronic states for the alloy have to lie between the bounds for the pure metals [13]. One can then analyze schematically the effect of chemical ordering in the following way. In the case of phase separation, the alloy density is the average of those of the pure metals whereas the subbands are narrower in the case of perfect order since the number of neighbours of the same type is reduced. For disordered systems, the width is in between but tails are present due to the finite probality of finding pure A and B clusters of any size. Finally, the effect of concentration is to reduce the width by reducing the number of neighbours. All these qualitative behaviours are illustrated in figure 8.

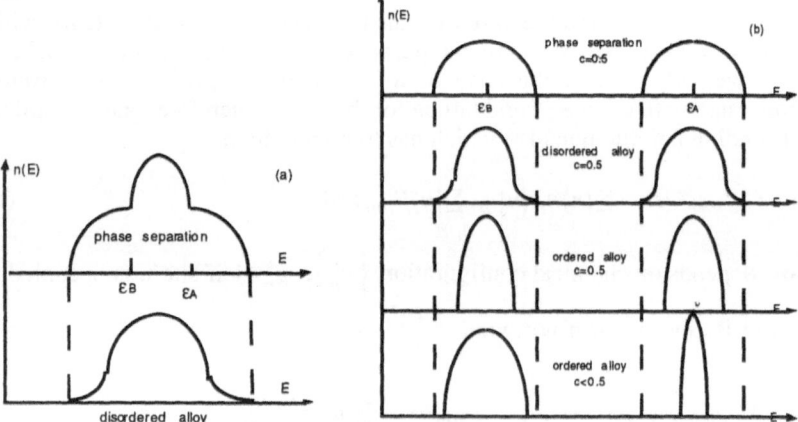

Figure 8: Schematic variation of the density of states upon alloying in the case of weak (a) and strong (b) diagonal disorder [13].

Let us now detail how to calculate more precisely the density of states for an alloy. In the ordered case, the same methods (continued fraction, recursion) can be used, taking just into account the ordered configuration of A and B atoms to assign the levels ε^A or ε^B. The situation is more complicated for a disordered system since it requires to calculate the average value, over all configurations, of n(E) and therefore G(E). The tight-binding hamiltonian can be re-written:

$$H = H^0 + V \tag{36}$$

$$H^0 = \sum_{n,m} |n\rangle \beta_{nm} \langle m| \quad \text{(pure metal)} \qquad\qquad V = \sum_n |n\rangle \varepsilon_n \langle n|$$

where only $\varepsilon_n = \sum_{i=A,B} p_n^i \varepsilon^i$ depends on concentration. If one notes $G^0 = (z-H^0)^{-1}$ the Green function for the pure metal, one can write a Dyson equation:

$G = (z-H^0-V)^{-1} = G^0 + G^0 V G^0 + G^0 V G^0 V G^0 + ... = G^0 + G^0 V G$

which leads to:

$\langle G \rangle = G^0 + G^0 \langle V \rangle G^0 + G^0 \langle V G^0 V \rangle G^0 + ... \neq [z-H^0-\langle V \rangle]^{-1}$

since the average of the product is not the product of the averages. One can then define (mean field approximation) an effective local potential $\Sigma(z)$ such as:

$$\langle G(z) \rangle = \left(z - H^0 - \Sigma(z) \right)^{-1} \qquad \text{with} \qquad \Sigma(z) \approx \sum_n |n\rangle \sigma(z) \langle n| \tag{37}$$

which means that in the average medium, the levels ε^A and ε^B are replaced by $\sigma(z)$ at each site. This effective potential can be determined by a self consistency condition which imposes that fixing the occupancy of a site and then making the average on this site would lead to recover the same potential. This is the *Coherent Potential Approximation* (CPA) [17], which leads to the condition:

$$\sum_i c^i t^i = 0 \qquad \text{with} \qquad t^i = \frac{\varepsilon^i - \sigma}{1 - \langle G_{NN} \rangle \left(\varepsilon^i - \sigma \right)} \tag{38}$$

which is indeed self-consistent since the Green function in the disordered state $\overline{G}(z) = \langle G_{NN}(z) \rangle$ depends on $\sigma(z)$. Since Σ is site-diagonal, it writes:

$$\overline{G}(z) = \langle G_{NN}(z) \rangle = G^0(z - \sigma(z)) \tag{39}$$

The alloy densities of states obtained in this way (recursion method for ordered system, CPA+continued fraction for disordered ones) [18] are in good agreement with those obtained by LMTO calculations [19].

2.2.2 Mixing energy

As previously stated for the pure metals the total energy of the alloy, for a given configuration, cannot be described as a sum of pair interactions. Nevertheless, the (small) part of the energy which depends explicitly on configuration (and which is essential in ordering problems) can be written as a sum of effective pair interactions in the following way, by developing the energy in a pertubative way with respect to the disordered state [20]:

$$E_{coh}\left(\left\{p_n^i\right\}\right) = \overline{E}(c) + \frac{1}{2} \sum_{n,m,i,j} p_n^i p_m^j V_{nm}^{ij} \tag{40}$$

$$V_{nm}^{ij} = -\frac{Im}{\pi} \int^{E_F} dE t_n^i(E) t_m^j(E) \sum_{\lambda\mu} \overline{G}_{nm}^{\lambda\mu}(E) \overline{G}_{mn}^{\mu\lambda}(E) \tag{41}$$

with the interatomic Green function: $\overline{G}_{nm}^{\lambda\mu}(E) = \langle n\lambda | \overline{G}(E) | m\mu \rangle$

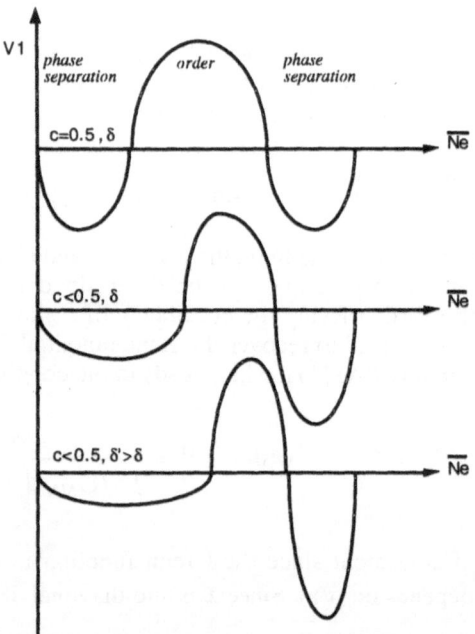

Figure 9: Variation of 1st neighbour effective pair interactions (42) for fcc bulk sites as a function of concentration and diagonal disorder.

Any energy balance which accounts for changes in the chemical configuration (mixing or ordering energies) will in fact involve the combination:

$$V_{nm} = \frac{\left(V_{nm}^{AA} + V_{nm}^{BB} - 2V_{nm}^{AB}\right)}{2} \tag{42}$$

the sign of which indicates the tendency of the system under study to order (V>0) or phase-separate (V<0). Moreover, one can show that V_{nm} decreases rapidly with the distance (n-m) ($V_1 \gg V_2$, V_3, $V_4 \gg V_5$, ... for fcc structure) and that it depends on bulk concentration (which could change the tendency to order or phase

separate in a system) and on the average d band filling $\overline{N}_e = cN_e^A + (1-c)N_e^B$
[20]. One can see in figure 9 that alloys with a nearly half filled band tend to order whereas those with a nearly filled or empty bands tend to phase separate

3. ELECTRONIC STRUCTURE OF SURFACES

3.1 Pure metals

3.1.1 Electronic structure

3.1.1.1 Normal metals: surface states: Let us consider a surface perpendicular to a reciprocal space vector $+/-G_0$, defining two edges of the Brillouin zone and directed along the z axis ($G_{0z}=|G_0|$). The surface plane is located at z_0 and the semi-infinite solide beyond: $z>z_0$. In the nearly free electron model, we have seen that the introduction of the lattice potential introduces a gap associated to G_0 (in $k_0=G_0/2$), corresponding to the requirement for k_0 to be real. The presence of the surface changes this condition. Indeed if one considers a state with momentum k_0 of components $k_{//}$ and k_z, such as $Re(k_z)=-G_{0z}/2$, one can now consider solutions of the Schrödinger equation inside the bulk gap since k_z can have an imaginary part k_I ($k_z=-G_{0z}/2+i\ k_I$), which means that one considers a wave which is damped in the solid. This introduces a condition on the energy for having a non trivial solution of the system. For each value of k_I which satisfies this condition, the Schrödinger equation has two real solutions for E, degenerate with those for $-k_I$. These dispersion curves are complex and connect the two sides of the gap by a loop with a real energy but complex momentum. The corresponding wave function ψ_{in} inside the solid is then damped. If the surface interrupts the lattice potential by a barrier at z_0, the Shrödinger equation outside (for $z<z_0$) has for solution an evanescent wave: ψ_{out}. A surface state is then a state with a wave function which exponentially decays on both sides of the surface. It only exists if one can match continuously the evanescent solution in vacuum ψ_{out} to the damped one in the solide ψ_{in}. This requires to satisfy a condition which depends on the sign of V_{G_0} [21].

3.1.1.2 Transition metals: if one neglects α, two bulk parameters should vary at the surface: first the hopping integrals $\beta(R_0=1^{st}$ neighbours) that we will assume unchanged at the surface (no relaxation), then the effective d level ε_d. Let us first assume that the latter is also unchanged at the surface. If one models again the local density of states by the schematic rectangular one with the same second moment as the exact one, it has a band width $W = dd\beta\sqrt{12Z}$ for a bulk site and

$W_s = dd\beta\sqrt{12Z_s}$ for a surface one with a reduced number of neighbours: $Z_s=Z-\Delta Z$
(ΔZ: number of broken bonds). Therefore, once the Fermi level is fixed by the bulk density of states, one finds an electronic charge transfer at the surface δN_s^0:

$$\delta N_s^0 = N_s^0 - N_e = \left(\sqrt{\frac{Z}{Z_s}} - 1 \right)(N_e - 5)$$

which means that there is a charge transfer $\delta N_s^0 \sim$ +(-) 1 electron at the end (beginning) of the transition series, which is unrealistic. Therefore the charge distribution must obey some self consistent rule. Indeed any modification of $n(E)$ at the surface $\delta n_s(E)$ induces a charge redistribution δN_s which in turn modify the potential and then ε_d. If one notes δV_s this modification of ε_d and assumes that it depends linearly on δN_s: $\delta V_s = U \delta N_s$ (U: Coulomb interaction), the self-consistent relation between charge and level linearises if δV_s small , so that one finds [21]:

$$\delta N_s \approx \delta N_s^0 - n_s(E_F)\delta V_s$$

$$\delta N_s^0 = (\frac{1}{U} + n_s(E_F))\delta V_s \approx n_s(E_F)\delta V_s$$

$$\}\Longrightarrow \quad \delta N_s = 0$$

The charge self consistency reduces to a <u>local neutrality condition</u> which determines the level modification δV_s :

$$\delta V_s = \frac{W}{10}\left(1 - \sqrt{\frac{Z_s}{Z}} \right)(N_e - 5) \approx \frac{W}{20}\frac{\Delta Z}{Z}(N_e - 5) \qquad (43)$$

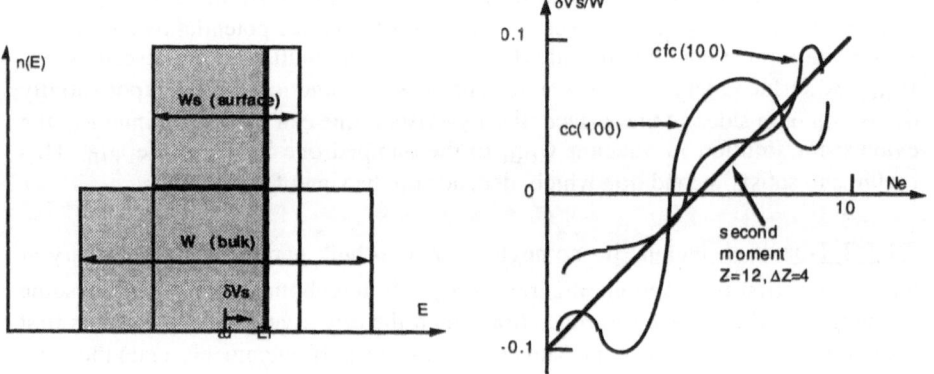

Figure 10: Variation with the d band filling (for realistic and rectangular densities of states) of the surface valence level shifts (right part) induced by the local charge neutrality requirement (schematized on the left part) [22].

This d level shift is almost rigidly followed by the core levels, which is confirmed experimentally by core level spectroscopy [22]. It follows some general trends which are illustrated in figure 10:

- $|\delta V_s|^{max} \sim W/10$

- δV_S changes sign in the middle of the series (which is slightly shifted for realistic densities: between W and Ta in the 5d series)
- $|\delta V_S|$ increases with the number of broken bonds (fcc: $|\delta V_{110}|>|\delta V_{100}|>|\delta V_{111}|$)
- for open surfaces one has also to modify the level on the first underlayer.

Let us note that this charge neutrality condition at the surface is confirmed by *ab initio* calculations (see figure 11 for Pd). More precisely the LMTO calculation of the surface densities of states (figure 12) even shows that, if one takes into account the sp-d hybridization, charge neutrality has to be achieved, not only per inequivalent site, but also per orbital [3].

Of course if one is interested in the more detailed modifications of the densities of states, one has to go beyond the second moment. In that case, one observes that, not only the density width is reduced at the surface, but also quasi surface states appear, which vanish beyond the surface layer. As said before, it is difficult for fcc elements of the end of the transition series to limit the description to pure d sates. It is necessary to introduce sp-d hybridization, which is possible in the tight-binding framework. The results, which strongly depend on the charge neutrality assumption (global or per orbital), are shown here for two orientations of a Pd surface (fig. 13) and compare satisfactorily to those of LMTO calculations (fig. 12). Moreover, sp-d hybridization leads to surface tensions in better agreement with experiments [11].

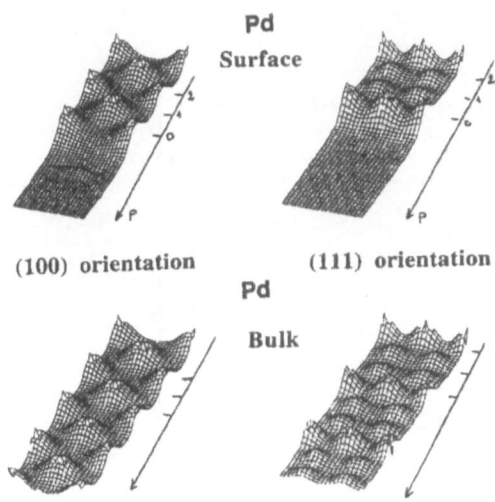

Figure 11: Charge distribution along the (100) and (111) directions for Pd with and without surface, from LMTO calculations [3].

G. Tréglia

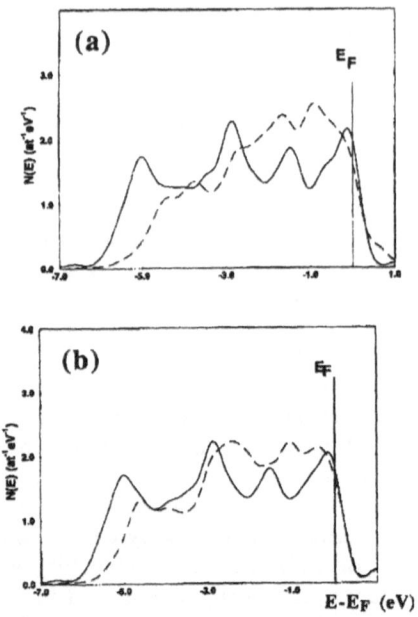

Figure 12: LMTO [3] surface densities of states (dotted line) for Pd (100) (a) and Pd (111) (b) compared to the bulk one (full line).

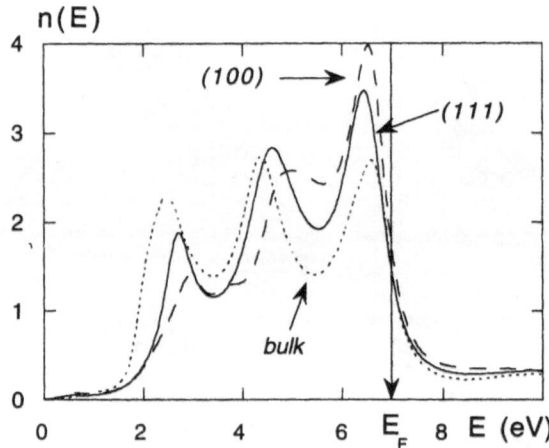

Figure 13: Surface density of states for Pd (100) and Pd (111) from tight binding calculations with sp-d hybridization (10 coefficients) [23].

3.1.2 Surface energy

The surface energy is defined as the energy which is required (per surface atom) to cut an infinite crystal in two parts.

3.1.2.1 Normal metals: for an infinite barrier, the creation of two surfaces confines the electrons inside since the wave function has to vanish at the surface. Then the limit conditions are no longer the same perpendicular and parallel to the surface, which changes the wave function. One state is missing at the bottom of the sub-band $k_{//}$ which has to be replaced at the Fermi level so that k_F is displaced. The electronic density is written:

$$\rho^-(z) = \sum_{k \leq k_F} \psi_k^* \psi_k \approx \rho_0^- \left(1 + 3 \frac{\cos(2k_F z)}{(2k_F z)^2} \right) \tag{44}$$

which leads to a charge oscillation which decreases along z. There is an excess of positive charges below the surface, due to the infinite barrier. In fact, if one reintroduces a finite barrier, the electrons can go out, leading to a surface dipole. Moving half of the electrons from $k_z=0$ to the Fermi level requires energy: the surface tension which is found to vary as $(k_F)^4$ [21].

3.1.2.2 Transition metals: in that case the creation of the surface induces a variation of the energy (9):

$$E_s = \delta E_{band} + \delta E_{rep} \tag{45}$$

$$\delta E_{band} = \int^{E_F} E \delta n(E, \delta V_S) dE - \frac{1}{2} \delta \left(\int^{E_F} d\vec{r} V^H(\vec{r}) n(\vec{r}) \right) \tag{46}$$

since the surface charge rearrangement $\delta n(r)$ modifies the Coulomb interactions between electrons, leading to a variation of Hartree potential δV_S and then again... up to self-consistency. This leads to:

$$\delta E_{band} = \int^{E_F} E \delta n(E, \delta V_s) dE - N_e \delta V_s - \left(\frac{1}{2} \delta N_e \delta V_s \right) \tag{47}$$

the last term vanishing due to the charge neutrality requirement ($\delta N_e=0$). The calculation in the second moment approximation is easy and leads to [21]:

$$\delta E_{band} = \frac{W_s - W}{20} N_e(10 - N_e) \qquad \delta E_{rep} = -\frac{q}{p} \frac{\Delta Z}{Z} W N_e(10 - N_e)$$

so that, up to first order with repect to broken bonds ΔZ :

$$E_s = \frac{p - 2q}{2(p - q)} \frac{\Delta Z}{Z} E_{coh} \tag{48}$$

which has a parabolic variation with N_e This is confirmed with more realistic densities of states, and consistent with experimental trends (the local minimum in the first series, which can be seen in figure 14, is attributed to correlation effects).

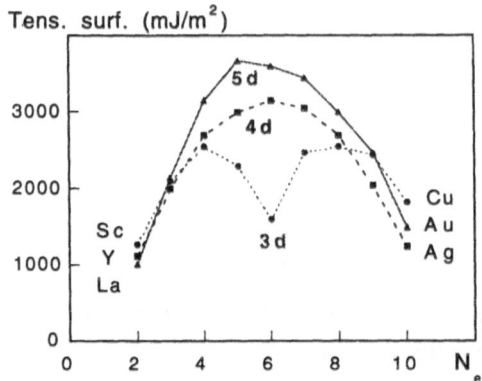

Figure 14: Experimental variation of surface tension along the transition series.

3.1.3 Atomic structure (transition metals)

3.1.3.1 Surface relaxation: let us now relax the first interlayer distance d_1 with respect to its bulk value (d_0). The surface energy (48) becomes a function of d_1 which involves the first two layers separated by d_1. The equilibrium value of d_1 is that which minimizes this energy. Up to first order with respect to d_1, one finds [21]:

$$\frac{\partial E_s}{\partial X} = 0 \quad ===> \quad \frac{d_1 - d_0}{d_0} = -\frac{\Delta Z}{4Z(p - 2q)} < 0 \tag{49}$$

This means that we find a contraction of the first interlayer spacing (inwards relaxation), of a few percents and proportional to the number of broken bonds (ΔZ). This is in agreement with experimental observations, contrary to simple pair potential models which predict an outwards relaxation. This result is due to the more rapid decrease with Z of the repulsive term compared to the attractive one ($\sim\sqrt{Z}$).

3.1.3.2 <u>Surface reconstruction</u>: it is less easy to analyze in a simple way the possible lateral atomic rearrangements since they involve both increasing and decreasing distances. It is then necessary to go more in details in the description of the cohesion. Nevertheless, one can see some trends which can be either along the transition series (zig-zag reconstruction of the (100) face of cc crystals) or along a column (missing row reconstruction of fcc (110) ones). These two reconstructions are illustrated in figure 15 and the corresponding systematics in the following table where the elements which undergo the reconstructions appear in bold characters.

V	**Cr**	...	...		*Ni*	*Cu*
Nb	**Mo**	...	...	*Rh*	*Pd*	*Ag*
Ta	**W**	...	...	**Ir**	**Pt**	**Au**

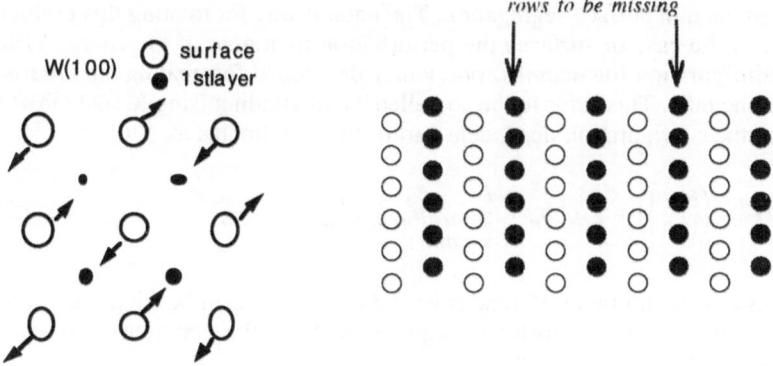

Figure 15: Zig-zag (cc(100)) and missing row (fcc (110)) reconstructions.

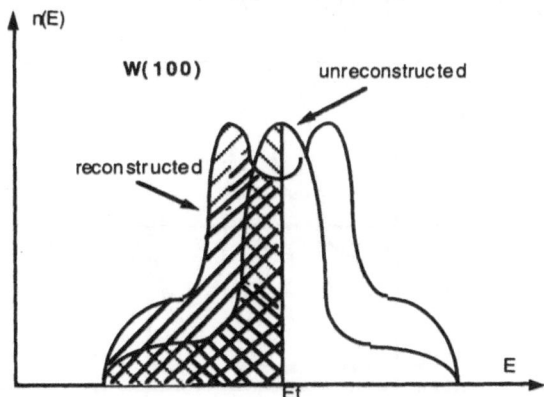

Figure 16: Influence of the zig-zag reconstruction on the (100) density of states of cc metals.

Both trends are well interpreted in the framework of tight-binding calculations. The zig-zag reconstruction is due to the broadening of the quasi-atomic surface peak of the local cc (100) density of states under the lattice distortion (see figure 16), which leads to an energy gain for d band filling around 5 (middle of the series) [24]. The missing row reconstruction of the (110) surface is attributed to the variation of the q parameter which drives the distance dependence of the hopping integrals: q_{Cu} $<q_{Ag} <q_{Au}$ [25].

3.2 Bimetallic systems

3.2.1 Alloy surfaces: $A_c B_{1-c}$

In the case of an alloy, the presence of the surface not only introduces atomic but also chemical rearrangements. Indeed, due to broken bonds, the equilibrium concentration at the surface has no reason to be the same as in the bulk, which leads to the phenomenon of surface segregation. The natural way for treating this problem is to extend to the case of surfaces the perturbation treatment of the energy (with respect to configuration fluctuations) previously developed for treating the ordering processes in the bulk. This leads to the so-called Tight-Binding Ising Model (TBIM) which writes the configuration-dependent part of the hamiltonian as [26]:

$$H^{mix}\left(\left\{p_n^i\right\}\right) = \sum_{n,i} p_n^i h_n^i + \frac{1}{2} \sum_{n,m,i,j} p_n^i p_m^j V_{nm}^{ij} \tag{50}$$

This presents essentially two differences with the case of the bulk. First, due to the existence of sites which are no longer equivalent from the geometrical point of view, it appears a local on-site term:

$$h_n^i = \frac{Im}{\pi} \int^{E_F} dE \sum_\lambda log\left[1 - \left(\varepsilon^i - \sigma_n\right)\overline{G}_{nn}^{\lambda\lambda}(E)\right] \tag{51}$$

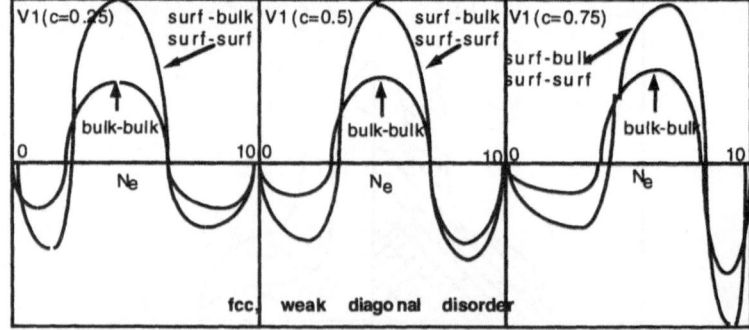

Figure 17: Enhancement of the effective pair interactions between first neighbours at the surface of fcc metals.

Moreover, the effective pair interactions which are given by (41-42) are enhanced at the surface with respect to the bulk, $1.5V < V_1 < 2V$, as can be seen in figure 17.

The concentration profile in the surface selvedge $\{c_p\}$, where $c_p=<p_n>$ (n : site in the pth plane parallel to the surface: p=0) is assumed to be homogeneous (mean field) and determined as the one which minimises the free energy:

$$F = \langle H \rangle - TS - \sum_p N_p \big(c_p - c\big)\mu \qquad (52)$$

where $\mu = \mu^A - \mu^B$ is the chemical potential. Denoting N_p the number of atoms in the p plane, each one having Z_p ($=Z^{tot}$ in the bulk) first neighbours among which Z_{pq} in the q plane, one finds:

$$\forall c_p, \quad \frac{\partial F}{\partial c_p} = 0 \quad ====> \quad \frac{c_p}{1-c_p} = \frac{c}{1-c} exp\left(-\frac{\Delta E_p^{TBIM}}{kT}\right) \qquad (53)$$

where the segregation energy is defined as:

$$\Delta E_p^{TBIM} = \Delta h_p + (1-2c)\, Z^{tot}\, V - \sum_{p'=-q}^{p'=+q}(1-2c_{p+p'})\, Z_{p,p+p'}\, V_{p,p+p'}$$

$$\Delta h_p = (h_p^A - h_p^B) - (h_{bulk}^A - h_{bulk}^B)$$

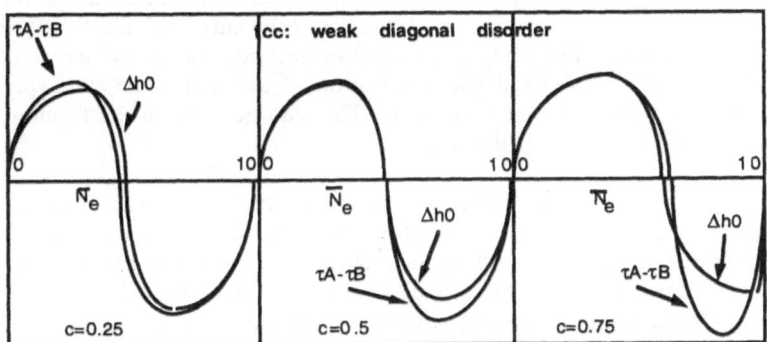

Figure 18: Comparisons between the variations with the d band filing of Δh_0 and $\tau^A - \tau^B$ for fcc metals.

The term Δh_p is numerically found almost identical to the difference in surface tensions $(\tau^A - \tau^B)$ for p=0 (see figure 18) and vanishes for p>0. Δh_0 is the major driving force and leads to the segregation of the element with the lowest surface tension. $V > (<) 0$ induces a segregation of the majority (minority) element and an oscillating (monotonous) profile.

Up to now, the derivation of TBIM has been performed on a rigid lattice, which is probably too crude in the case of large size mismatch between the constituents. However, it is possible to introduce this effect and that of the lattice relaxation by adding a third contribution to the segregation energy (53):

$$\Delta E_p = \Delta E_p^{TBIM} + \Delta H_p^{size}(c) \tag{55}$$

with $\Delta H_0^{size}(c) \neq 0$ if p=0 (1 for open surfaces). $\Delta H_0^{size}(c \text{-->} 0,1)$ is calculated in the framework of the second moment approximation, by determining the A-B parameters in order that A and B only differ by their size [14,27]. This leads to a contribution which significantly differs from that used in phenomenological approaches as derived from elasticity theory which led to a symmetric term with respect to (c,1-c), and then to the segregation of the impurity, whatever its size. On the contrary, the tight-binding term is found asymmetric with respect to (c,1-c), leading to a segregation of the impurity when it is the largest only (at least for close-packed surfaces). This comes from the anharmonicity of the potential which exhibits a strong asymmetry between tensile and compressive pressures.

It remains then to solve the non linear coupled equations (53) giving the concentration profile. The essential problem comes from that, due to this non linearity, many solutions can exist, and one has to be sure not only to get the one with the lowest energy (thermodynamic equilibrium) but also to find the metastable ones which could be stabilized under variation of external parameters (pression, temperature), leading to phase transitions. This can be done in the framework of phase portrait or local field methods [28].

Finally, let us mention some examples in which the use of this tight-binding model has been necessary to solve the segregation problem. In the case of PtRh [29] and PtPd [30] alloys, it has confirmed the local charge neutrality condition, invalidating the use of criteria based on electronegativity. For PtNi, for which $\Delta h_0 \sim 0$, the competition between the asymmetric size effect and the ordering one leads to profile phase transitions as a function of surface orientation and/or concentration and temperature [31], which have indeed been observed experimentally [32]. Finally, for a system with a strong tendency to phase separation such as CuAg, layering transitions occur [33], the first one having been observed experimentally [34].

3.2.2 Surface alloy: A/B

We will just mention that these energetic driving forces can also be included in a kinetical model, K(inetic-)TBIM, in order to describe the interdiffusion in the surface selvedge under annealing of a A metal deposited on a B substrate [35]:

$$\frac{dc_p}{dt} = f(c_p, c_{p-1}, \Delta H_p - \Delta H_{p-1}; c_p, c_{p+1}, \Delta H_p - \Delta H_{p+1}) \qquad (55)$$

The main results are the following. When the two elements present an ordering tendency (V>0), there is formation on a few layers of compounds A_cB_{1-c} similar to those of the bulk phase diagram if $\tau^A < \tau^B$, or unusual if $\tau^A > \tau^B$. On the contrary, if the elements tend to phase separate (V<0), new dissolution modes are predicted: "layer by layer" dissolution if $\tau^A < \tau^B$, "surfactant effect" if $\tau^A > \tau^B$ [36]. These *surface alloys* have been recently the subject of a lot of experimentally studies [37].

3.2.3 Atomic superstructure

In the case of the A/B deposit or of a strong segregation of the A element at the surface of a dilute B(A) alloy, one can expect reconstructions of the surface plane when the size-mismatch is large. The corresponding superstructures can be found by using Molecular Dynamics simulations in the tight-binding second moment approximation [38]. This is illustrated in figure 19 in the case of Ag/Cu (111), where a large ondulation is found at the surface, which seems to be confirmed by recent STM experiments [39].

Figure 19: Superstructure Ag / Cu (111) from quenched molecular dynamics study in a tight-binding potential [38].

3.2.4 Local order and densities of states

It is easy to calculate the average Green function and then the layer density of states, by extending the CPA to the case of inhomogeous systems (this requires to introduce different effective potentials σ_p for layers with different concentrations) [40]. The corresponding densities of states then depend on the local concentration, determined as explained above. They can then be calculated, in the framework of either tight-binding [41] or LMTO [42] calculations.

4. ELECTRONIC STRUCTURE OF CLUSTERS

4.1 Pure metals

4.1.1 Electronic structure

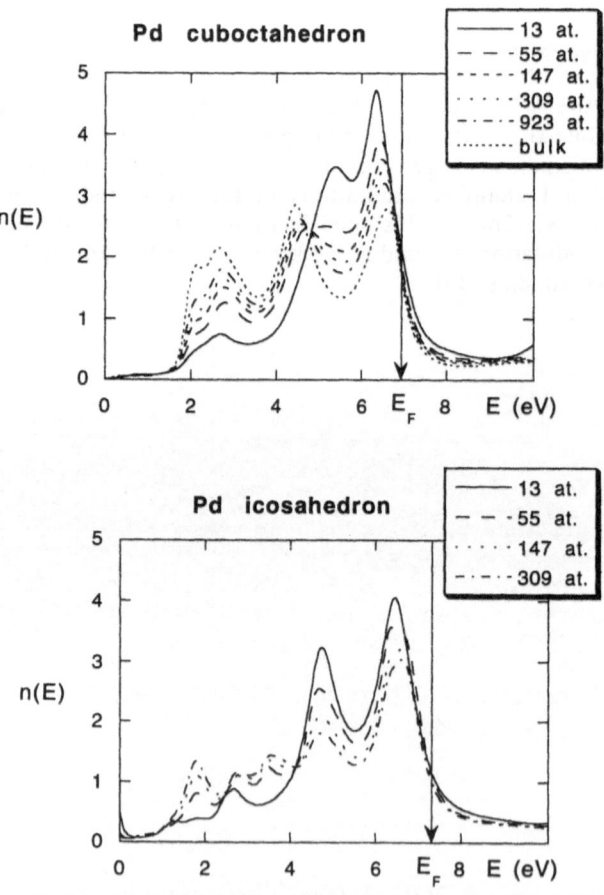

Figure 20: Size effect on the Pd cluster average density of states calculated in the tight-binding formalism, with an orbital neutrality rule [23].

Obviously, *ab initio* methods are particularly suited to the study of clusters with very small sizes but become very cumbersome when these sizes reach those which are useful for catalysis purposes (more than one hundred atoms). The tight-binding method is then very useful, since it describes the electronic structure in a wide range of sizes, and is able to give reliable site energies. Anyway, one has to take some caution. First, when interested in elements of the end of transition series (Pd, Pt, Cu), it is necessary to take into account the sp-d hybridization. Then, it is necessary to treat the relation between charge and potential, and the *ab initio* calculations have shown for the surfaces that it must involve a rule of charge neutrality which, in the case of sp-d hybridization, has to be achieved per orbital [3,23], which changes significantly the densities of states at the Fermi level with respect to those obtained using global site neutrality rules [11].

Doing that, the main results are that the band width decreases with the site coordination (from facets, to edges and vertices) and that the local densities are also significantly modified near the Fermi level depending on the site. Moreover the cluster symetry has a strong influence on the density of states (see figure 20). The influence of size and structure is illustrated in figure 20 where we plot the total density of states for cuboctahedral and icosahedral clusters, by taking the average of the densities on inequivalent sites (vertices, edges, facets, core) weighted by the corresponding number of sites [23].

4.1.2 Atomic structure

It is possible to obtain the equilibrium atomic configuration of the cluster at T=OK by performing a Quenched Molecular Dynamics study in the second moment interatomic potential. This leads to the relaxation profiles of figure 21 for Pd clusters with sizes between 13 (order n=1) and 4000 (order n=11) atoms [43]:

$$\delta R_p = \frac{R(p) - R_0(p)}{R_0(p)} \quad (>0: \text{means contraction of p shell})$$

where $R(p)$ $(R_0(p))$ is the radial distance of the pth shell to the center.

The most striking features which appear in figure 21 are that, both for cuboctahedron and icosahedron, one observes inhomogeneous contractions of the cluster and a curvature of the facets. Surprisingly, the icosahedron presents an "accordeon-like" profile, with a contraction both at the surface and in the core, leading to a large distance distribution around that of the bulk. It leads to a peculiar pression profile with a core tension. This tends to favour the formation of constitutive vacancies in the core, which stabilize clusters with central holes for a given size range [44]. Finally, the relaxation extends the stability range of icosahedron up to small sizes. The transition towards cuboctahedron appears at about 2000 atoms for Pd, lower for the Wulff polyhedron which is the stablest fcc cluster (the (100) facets are less developed). Finally, let us note that these results concerning the critical size for the transition strongly depend on the potentials [23].

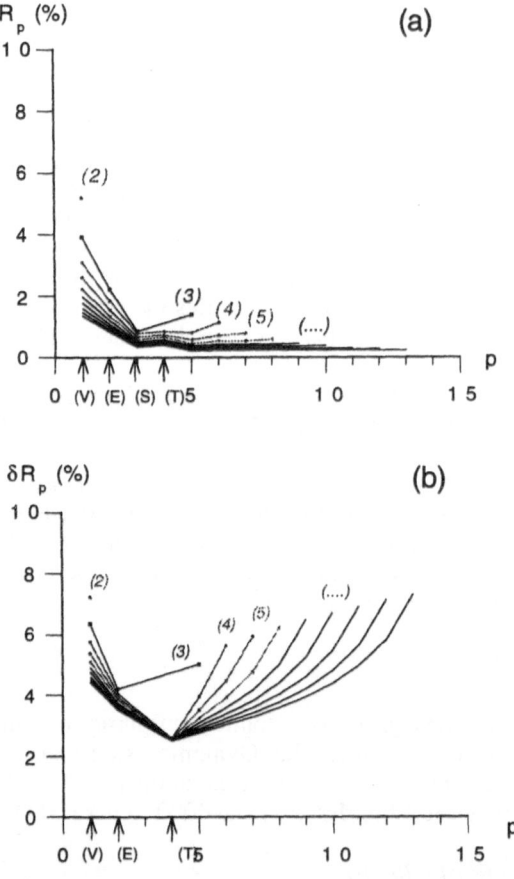

Figure 21: Radial relaxation profiles in Pd clusters, for cuboctahedron (a) and icosahedron (b) shapes [43].

4.2 Bimetallic systems

4.2.1 Finite size effect on surface segregation

Qualitatively, one expects that two effects will play an important role. First the effect of finite matter which means that, for dilute systems, the available quantity of segregant matter could be lower that the quantity of surface sites. This will occur for small sizes for which the ratio surface/bulk is high. Then, there may be a geometrical frustration due to the coexistence of facets with different orientations, which obliges to mix concentration profiles which could be antagonistic (ordering tendency), leading to antiphases boundaries. As a result the local order at the surface should be different from that of semi-bulk systems.

It is easy to extend TBIM to this finite case, by replacing layer concentrations by shells concentrations, the shell i being made of N_i sites. One can divide the surface shell into four subshells corresponding to vertices (i=1), edges (i=2), (100) and (111) facets (i=3,4). To simplify, let us first assume that the concentrations are homogeneous and only depend on coordination: $c_i=c_c$ for i>4. The concentration profile is then given by:

$$\frac{c_i}{1-c_i} = \frac{c_c}{1-c_c} \exp\left(-\frac{\Delta E_i(c_1,...,c_c)}{kT}\right) \qquad (56)$$

The effect of finite matter appears on the matter conservation rule, since now there is no longer an infinite reservoir. If one notes $D_i = \dfrac{N_i}{N_{tot}}$ the dispersion, this rule allows to determine c_c (..... and then c_i).

$$c_c\left(1 - \sum_{i=1}^{4} D_i\right) = c - \sum_{i=1}^{4} \frac{c_c D_i}{c_c + (1-c_c)\exp\left(\dfrac{\Delta E_i(c_1,...,c_c)}{kT}\right)} \qquad (57)$$

PtPd$_2$ (55 atoms)   **Pt$_3$Pd (55 atoms)**

Pt$_6$Pd$_4$ (923 atoms)   **Pt$_6$Pd$_4$(923 atoms)**

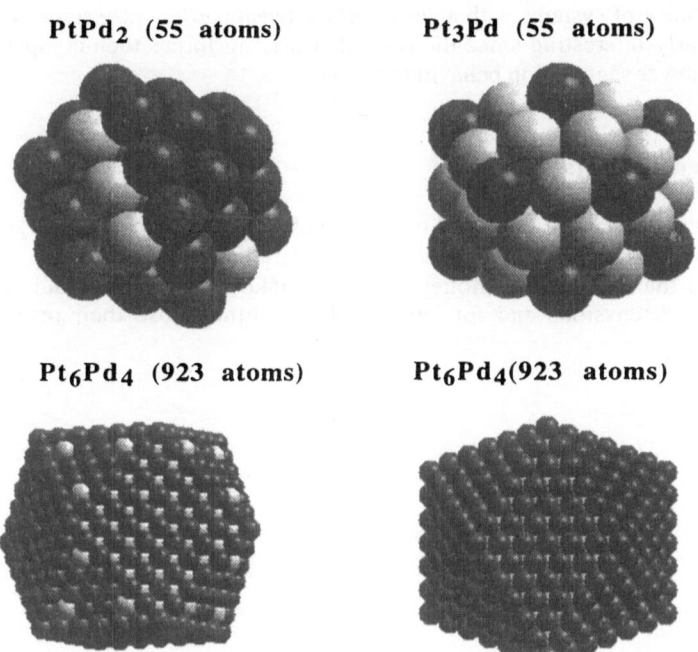

Figure 22: Competition between segregation and ordering in PtPd clusters (Pt atoms are white and Pd ones are black): cuboctahedron (left) and icosahedron (right) [30].

Moreover, there might exist a competition between ordering and surface tension effects. This is illustrated in figure 22 by performing Monte Carlo simulations in the TBIM potential for Pt-Pd clusters, at sufficiently low temperature (T=10K). Let us just recall that this system presents a tendency to ordering (V>0) and that Pd should segregate due to its lower surface tension ($\tau^{Pd} < \tau^{Pt}$). One sees for small clusters of 55 atoms and two concentrations (PtPd$_2$ cuboctahedron. Pt$_3$Pd icosahedron) that Pd first enriches the sites with low coordination (vertices, edges, ...). Then for larger clusters of 923 atoms with concentration Pt$_6$Pd$_4$ one sees that it occupies successively the vertices, then edges, square facets, but begins to occupy bulk sites before the triangular sites are filled, which allows to keep the L1$_0$ ordered structure. As a result, one finds many inequivalent surface sites, leading to some local order specific of the size and concentration.

4.2.2 Local densities of states and preferential sites

Once the equilibrium configuration is achieved, one can wonder what could be reactive sites for catalysis. One should then analyze the various local densities of states and their modifications with concentration and size, to see to what extent they are suited to the desired reaction. This should allow some "bimetallic catalyst design", by determining the characteristics (elements, concentration, size) suited to a given problem. From this point of view, if one recalls the case of semi-infinite PtNi, the choice of systems with a weak surface tension effect (see figure 14) should be particularly interesting since the two other driving forces then compete to give almost whatever segregation behaviour we want.

Acknowledgements

The author is particularly indebted to François Ducastelle and Bernard Legrand for their invaluable help and advices concerning the matter of this lecture. He would also like to thank Christine Mottet, Jacek Goniakowski, Andrés Saúl and Samir Sawaya for discussions and for having allowed him to use their results before publication.

References

[1] Hohenberg P. and Kohn W., *Phys. Rev.* **136 B** (1964) 864
     Kohn W. and Sham L., *Phys. Rev.* **140 A** (1965) 1133
[2] Skriver H.L., The LMTO Method, *Springer Series in Solid State Science* **41**
[3] Sawaya S., Goniakowski J., Mottet C., Saúl S. and Tréglia G., to appear
[4] Car R. and Parrinello M., *Phys. Rev. Lett.* **55** (1992) 2471
[5] Ashcroft N.W., *Philos. Mag.* **8** (1963) 2055
[6] Friedel J., Physics of Metals **1** (Cambridge University Press, 1978)
[7] Cyrot-Lackmann F., *Adv. Phys.* **16** (1967) 393
     Ducastelle F., *J. Physique* **31** (1970) 1055
     Cyrot-Lackmann F. and Ducastelle F., *Phys. Rev. Lett.* **27** (1971) 429
     Ducaselle F., thesis (Orsay, 1972)
[8] Gaspard J.P. and Cyrot-Lackmann F., *J. Phys.* C **6** (1973) 3077
[9] Haydock R., Heine V. and Kelly M.J., *J. Phys.* C **5** (1972) 2845
     Haydock R., Heine V. and Kelly M.J., *J. Phys.* C **8** (1975) 2591
[10] Turchi P., Ducastelle F. and Tréglia G., *J. Phys.* C 15 (1982) 2891
[11] Mottet C., Tréglia G. and Legrand B., *Surf. Sci.* **352-354** (1996) 675
[12] Heine V. and Weaire D., *Solid State Phys.* **24** (1970) 250
[13] Ducastelle F., Order and Phase Stability in Alloys, North-Holland, 1991
[14] Rosato V., Guillopé M. and Legrand B., *Philos. Mag. A* **59** (1989) 321
[15] Foiles S.M., Baskes M.I. and Daw M.S., *Phys. Rev. B* **33** (1986) 7983
[16] Garofalo M., Tosatti E. and Ercolessi F., Surf. Sci. 188 (1987) 321
[17] Velicky B., Kirkpatrick S. and Ehrenreich H., *Phys. Rev. B* **175** (1968) 747
[18] Bieber A., Ducastelle F., Gautier F., Tréglia G. and Turchi P., *Solid State Comm.* **45** (1983) 585
[19] Jarlborg T., Junod A. and Peter M., *Phys. Rev B* **27** (1983) 1558
     Kudrnovsky J., Bose S.K. and Andersen O.K., *Phys. Rev. B* **43** (1991) 4613
[20] Ducastelle F. and Gautier F., *J. Phys. F* **6** (1976) 2039
     Bieber A., Gautier F., Tréglia G. and Ducastelle F., *Solid State Comm.* **39** (1981) 149
[21] Desjonquères M.C. and Spanjaard D, Concepts in Surface Physiscs, Springer-Verlag, 1995
[22] Spanjaard D., Guillot C., Desjonquères M.C., Tréglia G. and Lecante J., *Surf. Sci. Rep.* **5** (1985) 1
[23] Mottet C., thesis (Marseille, 1996)
[24] Legrand B., Tréglia G., Desjonquères M.C. and Spanjaard D., *J. Phys. C* **19** (1986) 4463
[25] Guillopé M. and Legrand B., *Surf. Sci.* **215** (1989) 577
[26] Tréglia G., Legrand B. and Ducastelle F., *Europys. Lett.* **7** (1988) 575
[27] Tománek D., Aligia A.A. and Balseiro C.A., *Phys. Rev. B* **32** (1985) 5051
     Tréglia G. and Legrand B., *Phys. Rev. B* **35** (1987) 4338
[28] Ducastelle F., Legrand B. and Tréglia G., *Prog. Theor. Phys. Suppl.* **101** (1990) 159
[29] Legrand B. and Tréglia G., *Surf. Sci.* **236** (1990) 398
[30] Khoutami A., thesis (Orsay, 1993)
[31] Legrand B., Tréglia G. and Ducastelle F., *Phys. Rev. B* **41** (1990) 4422

[32] Gauthier Y. and Baudoing R., Surface Segregation and Related Phenomena,
      Boca Raton, CRC Press, 1990, p 169
[33] Saùl A., Legrand B. and Tréglia G., *Phys. Rev. B* **50** (1994) 1912
[34] Eugène J., Aufray B. and Cabané F., *Surf. Sci.* **273** (1992) 372
[35] Senhaji A., Tréglia G., Legrand B., Barrett N.T., Guillot C. and Villette B.,
      Surf. Sci. **274** (1992) 297
[36] Legrand B., Saùl A. and Tréglia G., *Mat. Sci. Forum* **155-156** (1994) 165
      Tréglia G., Legrand B. and Saùl A., *Il Vuoto* (1996) in press
[37] Bardi U.,*Rep. Prog. Phys.* **57** (1994) 939
[38] Mottet C., Tréglia G. and Legrand B., *Phys. Rev. B* **46** (1992) 16018
[39] Aufray B., Göthelid M., Gay J.M., Mottet C. and Landemark E., submitted to
      Surf. Sci. Lett.
[40] Berk N.F., *Surf. Sci.* **48** (1975) 289
[41] Lambin P. and Gaspard J.P., *J. Phys. F* **10** (1980) 651
      Lambin P. and Gaspard J.P., *J. Phys. F* **10** (1980) 2413
[42] Kudrnovsky J., Bose S.K. and Drchal V., *Phys. Rev. Lett.* **69** (1992) 308
      Ruqian Wu and Freeman A.J., *Phys. Rev. B* **52** (1995) 12419
[43] Khoutami A., Legrand B., Mottet C. and Tréglia G., *Surf. Sci.* **307-309**
      (1994) 735
[43] Mottet C., Tréglia G. and Legrand B., submitted to *Surf. Sci. Lett.*

Étude des adsorbats moléculaires sur surface métallique par spectroscopies de photoémission et d'absorption X

G. Tourillon

Laboratoire de Cristallographie-CNRS, 25 boulevard des Martyrs, B.P. 166, 38000 Grenoble cedex 09, France

---

Le champ d'applications des polymères, des molécules organiques et de leurs interfaces avec des métaux est si vaste que ces matériaux sont devenus indispensables dans la plupart des technologies modernes. Les polymères, en effet, se substituent très souvent aux métaux même lorsque les conditions d'utilisation sont astreignantes (températures élevées, contraintes mécaniques...). Ces matériaux sont également de plus en plus utilisés dans les domaines de la corrosion, de la catalyse, de la lubrification, de l'emballage alimentaire ou comme biomatériaux.

Cependant le développement de toutes ces applications suppose une meilleure connaissance des relations entre propriétés macroscopiques de ces matériaux, de leurs interfaces et de leurs structures. Bien que de nombreuses études aient déjà été réalisées, ces systèmes et tout particulièrement leurs interfaces avec un métal sont encore mal caractérisés, ce qui nécessite des efforts particuliers pour préciser :

1) la nature des interactions ;
2) le type de liaisons entre le substrat et le matériau ;
3) la composition chimique et structurale des premières couches déposées sur la surface métallique ;
4) l'organisation cristallographique et l'orientation de ces premières couches sur la surface.

Ces recherches imposent donc d'utiliser et de développer des méthodes spectroscopiques permettant de remonter in-situ aux propriétés physico-chimiques de ces systèmes. Il est par ailleurs important d'en avoir une approche multitechnique afin d'être capable de déterminer précisément les propriétés électroniques et structurales de ce type d'interfaces.

L'adsorption se définit comme la probabilité de rencontre d'une molécule organique avec la surface métallique, cette probabilité dépendant de la température et de la pression[1]. Dans un premier temps, cette molécule va être attirée par la surface car son dipôle subit son image électrostatique au travers du métal (effet de la force image). Par contre, lorsqu'elle est suffisamment proche, la répulsion coulombienne due à l'interaction de son nuage électronique avec les électrons du métal devient prépondérante.

Si la température n'est pas suffisante, la molécule restera en équilibre au-dessus de la surface sans recouvrement de ses orbitales avec celles du métal : c'est **la physisorption**.

Par contre si elle est capable de franchir la barrière coulombienne, ses orbitales vont s'hybrider avec les bandes du métal conduisant à la **chimisorption**.

Il s'agit donc de déterminer les paramètres qui gouvernent les mécanismes d'interactions entre molécule et métal : rôle de la structure chimique, propriétés électroniques du métal, surface cristallographique... Dans cette optique, les spectroscopies de photoémission de coeur (XPS), de valence (UPS), d'absorption X (NEXAFS), de déexcitation (Auger résonant et Raman résonant) et d'infra-rouge sont tout à fait adaptées et complémentaires pour étudier cette chimie de surface.

1. SPECTROSCOPIE DE PHOTOÉMISSION XPS ET UPS

La photoémission est basée sur l'effet photoélectrique c'est-à-dire que lorsqu'un électron absorbe un photon d'énergie égale ou supérieure à son énergie de liaison, il est éjecté de son niveau orbitalaire avec une certaine énergie cinétique. Le trou ainsi créé va induire une réorganisation des autres électrons situés sur des niveaux énergétiques moins profonds. D'autre part, le photoélectron subit l'influence coulombienne de ce trou. Il en résulte plusieurs mécanismes de relaxation[2-6] :

- Si l'électron est rapidement éjecté (**limite soudaine**), les orbitales se contractent brutalement pour écranter le trou, ce qui va conduire à l'apparition de "shake up" (transitions multiélectroniques entre des niveaux de valence et des niveaux discrets vides) ou/et de "shake off" (transitions dans le continuum) ;

- Par contre, si cet électron traverse lentement le nuage électronique (**limite adiabatique**), le trou reste écranté suffisamment longtemps par le photoélectron pour que les autres électrons relaxent d'une manière adiabatique. Dans ces conditions, les phénomènes de shake up ou off ne sont pas observés.

L'**XPS** utilise généralement des énergies de photons de ~1.2 ou 1.4 keV et se traite souvent dans l'approximation soudaine. Cette spectroscopie qui sonde les niveaux de coeur localisés permet de remonter à la **nature chimique des éléments** présents dans la matière.

L'**UPS** qui est associé à des énergies de photons de quelques dizaines d'eV permet d'étudier les **densités d'états occupés**. Elle est très souvent décrite dans l'approximation adiabatique et ne présente pas le caractère sélectif de l'XPS puisqu'elle sonde des niveaux de valence ou orbitalaires délocalisés.

Ces deux techniques qui étaient initialement des "méthodes de laboratoire", font de plus en plus appel au rayonnement synchrotron car il est possible de choisir l'énergie d'excitation. La section efficace d'interaction et le libre parcours moyen des photoélectrons peuvent donc être optimisés de telle sorte que l'on sonde uniquement la très proche surface.

Lorsqu'une molécule organique est adsorbée sur une surface métallique, ses orbitales sont affectées ce qui induit des changements tant en position en énergie (liée à la nature chimique) qu'en largeur (durée de vie du trou connecté aux phénomènes d'hybridation) pour les pics de photoémission, comparés à ceux obtenus pour la même molécule en phase gaz ou en phase condensée. Ces spectroscopies donnent donc des renseignements sur les mécanismes d'interactions. Des phénomènes dynamiques d'écrantage métallique ou d'apparition de force image peuvent cependant se produire lors de l'éjection du photoélectron, conduisant également à des déplacements des niveaux de coeur. Ces modifications peuvent donc provenir soit de l'état initial lors de l'adsorption soit de l'état final ionique résultant de l'interaction du photon avec la matière. Il n'est pas toujours possible de séparer la contribution de ces deux phénomènes dans les spectres de photoémission ; d'autres techniques spectroscopiques sont alors nécessaires.

2. SPECTROSCOPIE D'ABSORPTION X-NEXAFS [7]

La spectroscopie d'absorption X aux seuils des éléments légers ou NEXAFS est comme la photoémission basée sur l'effet photoélectrique. La différence réside dans le fait que l'on utilise un faisceau monochromatique d'énergie variable (et non pas fixe comme en XPS ou en UPS) autour du seuil d'absorption de l'élément sélectionné (par exemple 285eV pour le seuil K du carbone).

Près du seuil d'absorption, le photoélectron est éjecté avec une énergie inférieure au potentiel d'ionisation et il va être "piégé" dans les **premiers états vides** liés (orbitales de type π^* pour les composés organiques) ; l'état final est donc neutre et le spectre NEXAFS sera caractérisé par des transitions intenses et étroites. Au-delà du potentiel d'ionisation, l'état final est ionique et l'extension spatiale des orbitales devient important ; le photoélectron peut donc migrer ce qui réduit le temps de vie de l'état excité et induit un élargissement considérable des transitions (très souvent de symétrie σ pour les molécules).

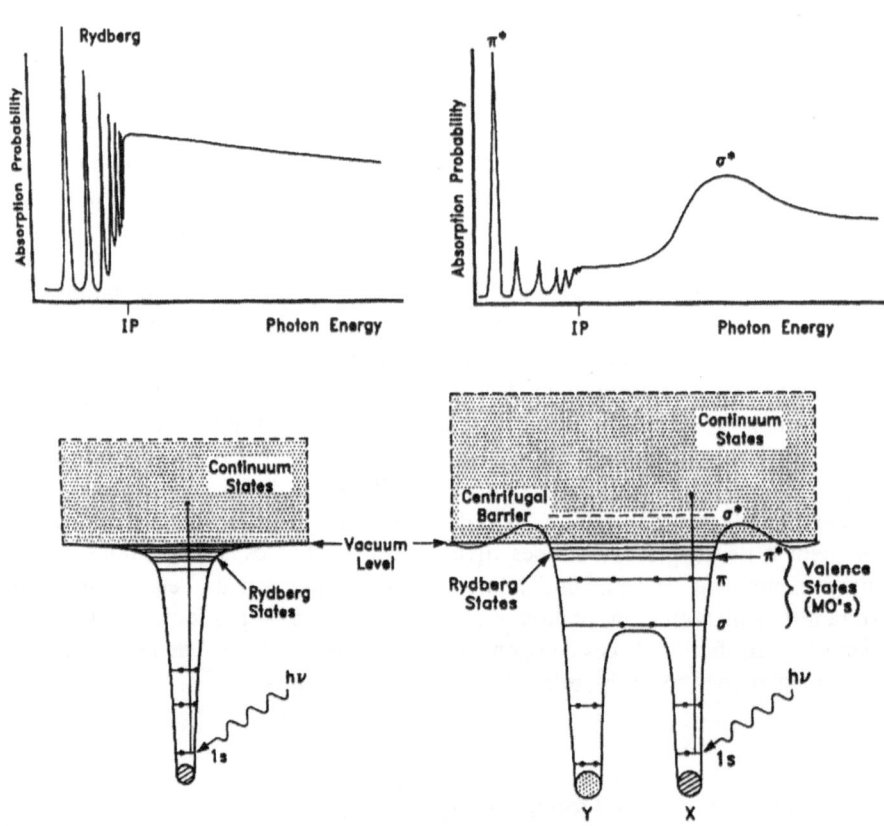

Fig. 1. — *Schéma de principe de la spectroscopie d'absorption X pour une phase gaz (figure de gauche) où seul est observé un seuil d'absorption X et des transitions Rydberg alors que pour la matière condensée (figure de droite) la présence de voisins autour de l'atome sélectionné entraîne des structures "fines".*

D'autre part, **la position en énergie de ces transitions sera caractéristique des liaisons intramoléculaires** puisque le photoélectron a une faible énergie cinétique et donc un libre parcours moyen important (phénomènes de diffusion multiple). Ceci entraîne la possibilité de détecter la présence de liaisons spécifiques (liaisons CC simples, doubles triples, CO,CN...) et de "mesurer" leurs longueurs grâce à la valeur énergétique.

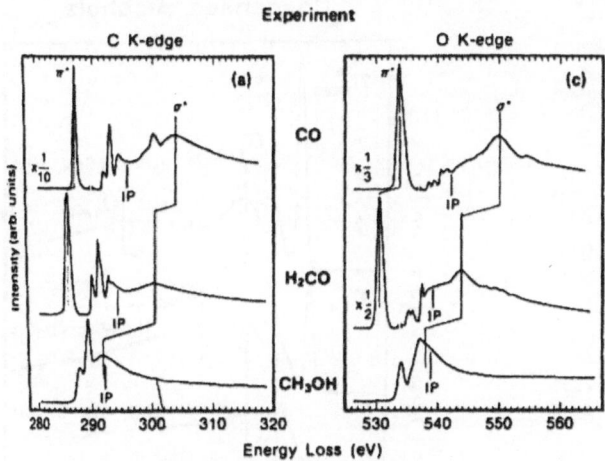

Fig. 2. — Spectres NEXAFS aux seuils Carbone et oxygène de diverses molécules organiques possédant une simple, une double ou une triple liaison C-O montrant le déplacement en énergie de la résonance de forme σ^* suivant la nature de cette liaison.

Pour des molécules complexes, le principe de "l'addition des blocs" a été introduit par Stohr, c'est-à-dire que le spectre NEXAFS d'une molécule composée de plusieurs groupements fonctionnels est constitué de la somme des structures caractéristiques de chaque fonction.

Cette notion n'est valable que si les orbitales anti liantes sont suffisamment localisées ce qui exclut de ce modèle les molécules conjuguées (ex du benzène, butadiène, acrylonitrile à π^* conjuguées ou les hydrocarbures saturés à longue chaîne aliphatique à σ^* conjuguées).

La spectroscopie d'absorption X fait obligatoirement appel au rayonnement synchrotron puisqu'elle nécessite une source d'excitation d'énergie variable et continue. De plus, **grâce à la polarisation linéaire de ce faisceau, il est possible de préciser l'orientation des molécules** par rapport à la surface métallique simplement en variant l'angle entre cette surface et le faisceau de photons.

La comparaison des spectres NEXAFS d'une molécule adsorbée en monouche ou en submonocouche sur une surface à ceux obtenus pour la même molécule en phase gaz ou en phase condensée permettra de préciser les modes d'interactions. En effet, pour les états liés en dessous du potentiel d'ionisation (typiquement les orbitales π^*), l'état final est neutre :

- Si la molécule est physisorbée, donc en faible interaction, aucune modification tant au niveau position en énergie qu'en intensité et largeur n'est attendue entre phase adsorbée et phase condensée (ou phase gaz).

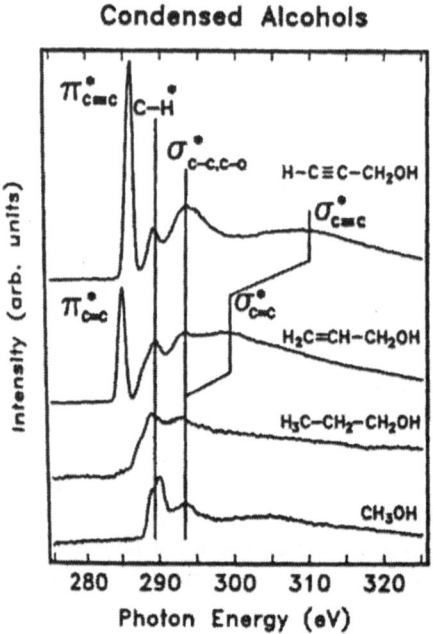

Fig. 3. — *Spectres NEXAFS de molécules organiques possédant diverses fonctions chimiques montrant l'approche du "building block picture" faite par Stohr.*[7]

- Par contre si la molécule est chimisorbée, l'hybridation de ses orbitales avec le métal provoque l'apparition de niveaux vides supplémentaires. Ces états ont un fort caractère métallique et le photoélectron ne sera plus localisé sur l'orbitale.

La durée de vie de l'état final diminue ce qui va provoquer un élargissement des transitions.

Le NEXAFS donnera donc des informations complémentaires à celles de la photoémission.

Les potentialités de cette méthode se sont récemment accrues avec le développement de monochromateurs ayant une résolution en énergie de ~20meV à 300eV, c'est-à-dire inférieure à la largeur naturelle du trou profond pour les éléments légers (typiquement 100-110 meV pour le carbone). Ces dispositifs permettent ainsi d'obtenir des informations sur les états vibrationnels des molécules à l'état excité (modifications de symétrie et d'angles de liaisons, modes de réarrangements)[8,9]. Ces ruptures de symétrie peuvent conduire à des modifications des orbitales moléculaires ce qui a des implications sur la réactivité de surface. De nouveau une comparaison avec la spectroscopie IR "classique", c'est-à-dire à l'état fondamental, est nécessaire.

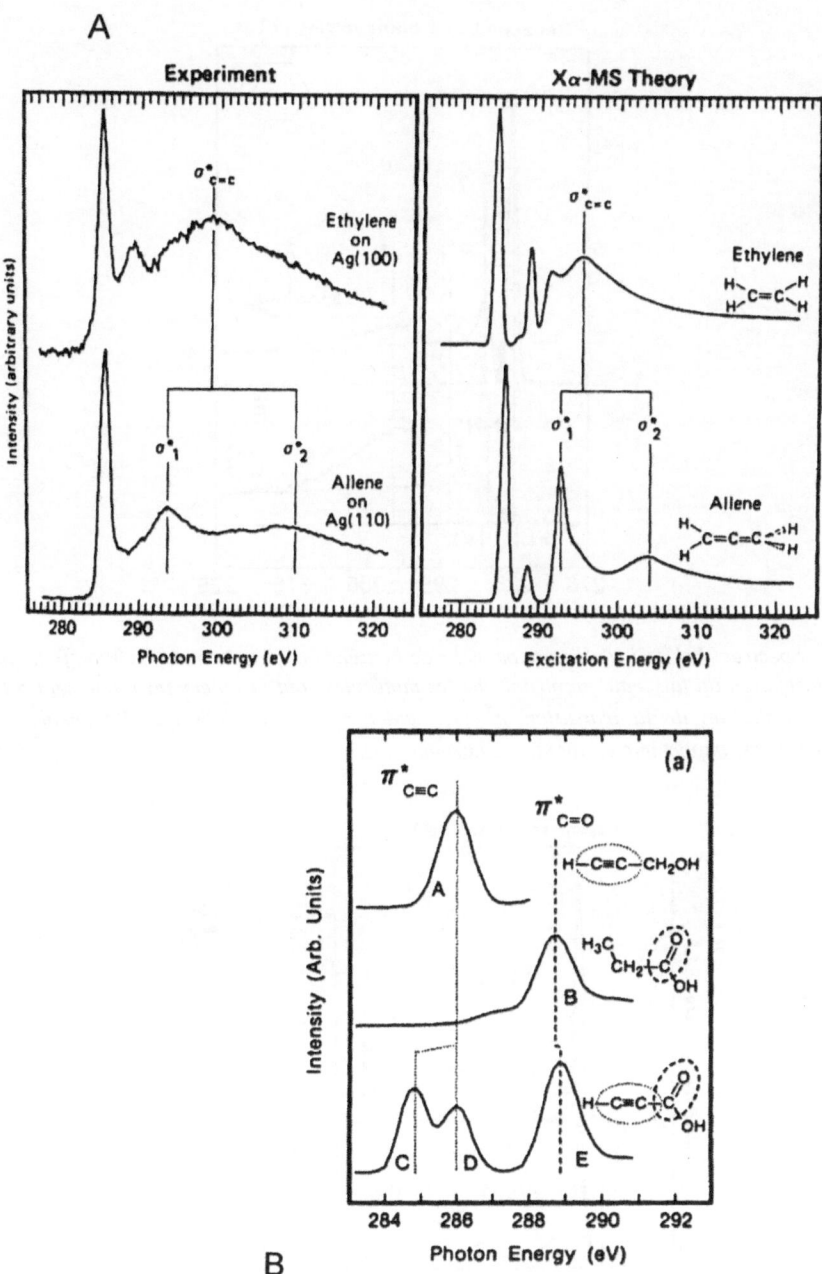

Fig. 4. — *Spectres NEXAFS montrant que le modèle du "building block picture" n'est pas valable lorsque des effets de conjugaison entre liaisons de symmetrie σ (A) ou π (B) existent dans les molécules organiques (apparition de "nouvelles" transitions dues à la conjugaison).*

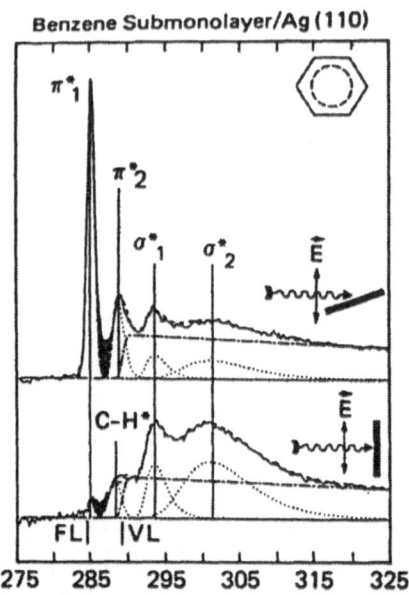

Fig. 5. — *Spectres NEXAFS d'une monocouche de benzène adsorbée sur Ag(110) en fonction de la polarisation du faisceau, montrant que les molécules sont orientées parallélement à la surface (exhaltation de la transition* π^* *en incidence rasante alors que l'intensité des transitions* σ^* *est importante en incidence normale).*

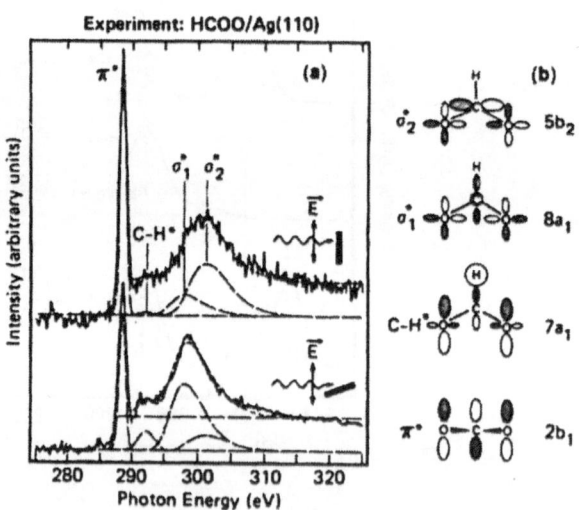

Fig. 6.— *Spectres NEXAFS au seuil du carbone d'une monocouche d'acide formique adsorbée sur Ag(110) à 95K qui révèlent la formation de formiate sur la surface (a), en accord avec les calculs d'orbitales moléculaires (b).*

3. SPECTROSCOPIE DE DÉEXCITATION (AUGER RÉSONANT ET RAMAN RÉSONANT) [8,10,11]

Les spectroscopies d'absorption X et de photoémission sont largement utilisées pour remonter aux mécanismes d'interaction de molécules organiques avec une surface métallique (modifications des orbitales moléculaires, apparition de nouveaux états résultant de l'hybridation...). Il est quelquefois difficile de mettre en évidence ces nouveaux états à cause de leur recouvrement avec la bande de valence du métal. La spectroscopie de déexcitation (Auger résonant lié aux électrons et Raman résonant avec les photons de fluorescence) s'avère être une méthode de choix pour séparer ces différentes contributions car elle est sélective du point de vue site atomique et symétrie des orbitales (pour le Raman résonant voir l'exposé de C.Hague).

Lorsqu'un électron de coeur est placé sur un niveau vide de symétrie π ou σ, les processus de déexcitation peuvent soit mettre en jeu cet électron excité (électron participateur) soit se produire suivant un "schéma classique" Auger modifié par la présence de cet électron sur le niveau vide (électron spectateur).

Le processus participateur conduit à un état final ayant un seul trou sur une orbitale de valence de la molécule ; il est donc similaire à celui de la photoémission. Cependant l'intensité des transitions sera différente car dans un cas (participateur), les règles de sélection sont gouvernées par les éléments de matrices Auger alors que dans l'autre cas (photoémission), elles dépendent des éléments de matrices dipolaires. L'autre différence vient également du fait que ce processus a un **caractère très localisé** (matrices Auger) ; il sera donc possible, en excitant sélectivement un site atomique dans une molécule, de sonder les fonctions d'onde des différentes orbitales moléculaires à partir de ce site.

Le processus spectateur conduit par contre à un état final composé de deux trous sur les orbitales de valence et un électron sur une orbitale antiliante. Il nous donnera donc des renseignements sur les modes d'écrantage des trous par cet électron sur l'orbitale antiliante.

Pour des molécules chimisorbées et donc en fortes interactions avec le substrat, des transferts de charge peuvent se produire entre l'excitation et la déexcitation (écrantage du trou de coeur par transfert de charge du métal vers la molécule). Dans ces conditions, le spectre de déexcitation doit être similaire au spectre Auger normal (renseignement sur les phénomènes **dynamiques de transfert de charge**). Enfin, l'électron placé sur une orbitale antiliante peut migrer et ne plus être présent au moment de la déexcitation ce qui permet d'accéder au **temps de vie de cet état électronique**.

L'expérience consiste à enregistrer tout d'abord un spectre de photoémission pour une énergie d'excitation légèrement inférieure au seuil d'absorption de l'élément choisi (par exemple 270eV pour le carbone ou 390eV pour l'azote) puis un spectre Auger normal à une énergie d'excitation nettement supérieure au seuil d'absorption (320eV pour C ou 430eV pour N) et enfin les différents spectres Auger résonant aux énergies d'excitation correspondantes aux transitions observées dans le spectre d'absorption X.

En conclusion, l'étude des mécanismes d'interactions qui se produisent lors de l'adsorption d'une molécule organique sur une surface métallique nécessite une approche multispectroscopique si l'on veut décrire précisément les processus de transfert de charge. L'exemple de l'adsorption de CO sur Ni(100), discuté en détail dans la littérature[10,12,13], montre l'apport de chaque spectroscopie à la compréhension des mécanismes de transfert de charge à l'interface.

Références

[1] N.H. March, "Chemical bonds outside metal surfaces", Plenum Press, New York (1986).
[2] H. Tillborg, A. Nilsson and N. Martensson, *J. Electron. Spectrosc. Relat. Phenom.* **62** (1993) 73.
[3] A.T. Amos, B.L. Burrows and S.G. Davison, *Surf. Sci.* **277** (1992) L100.
[4] W. Wurth, D. Coulman, A. Puschmann, D. Menzel and E. Umbach, *Phys.Rev.B* **41** (1990) 12933.
[5] D. Nordfors, H. Agren and K.V.Mikkelsen, *Chem.Phys.* **164** (1992) 173.
[6] M.C. Desjonquères and D. Spanjaard "Concept in Surface Physics" Springer Series in Surface Science, Springer Verlag, Berlin Vol. **30** (1992) 422.
[7] J. Stohr "NEXAFS Spectroscopy", Springer Series in Surface Science, Springer Verlag, Berlin, Vol.**25** (1992).
[8] N. Martensson and A. Nilsson in "Applications of Synchrotron Radiation-High resolution studies of molecules and molecular adsorbates on surfaces", W. Eberhardt Ed., Springer Series in Surface Science, Springer Verlag,Berlin, Vol. **35** (1994).
[9] M.P. de Miranda, A. Beswick, P. Parent, C. Laffon, G. Tourillon, A. Cassuto, G. Nicolas and F.X. Gadea, *J.Chem.Phys.* **101** (1994) 5500.
[10] Adsorbate core hole excitation and decay dynamics, O. Bjorneholm, Thesis, Uppsala, Sweden (1992).
[11] T. Aberg and B. Craseman in "Resonant Anomalous X-ray Scattering : Theory and Applications", G. Materlik, C.J. Sparks and K.Fischer Eds., Elsevier Science (1994).
[12] A.Nilsson, *Phys.Rev.B* **51** (1995) 10224.
[13] R.Hoffmann, *Rev.Mod.Phys.* **60** (1988) 601.

Catalysis by Metals: Contribution of Electrochemistry

J. Barbier, M.J. Chollier and F. Epron

Laboratoire de Catalyse en Chimie Organique, URA 350 du CNRS, Université de Poitiers, 40 avenue du Recteur Pineau, 86022 Poitiers, France

---

In Catalysis by metals, methods of electrochemistry can be mainly used in the two following areas :
- Preparation of bimetallic catalysts by surface redox reactions,
- Characterization of metallic catalysts.

1. PREPARATION OF BIMETALLIC CATALYSTS BY SURFACE ELECTROCHEMICAL REACTIONS

Preparation procedures of bimetallic catalysts influence the type of interaction between the two metallic species and then the catalytic performance of catalysts.

Different electrochemical methods can be used to bring into close contact the two metals responsible for the formation of the bimetallic entities. Four methods of preparing bimetallic catalysts were developed. These methods are direct redox reactions, redox reactions of adsorbed species, catalytic reduction and underpotential deposition [1].

1.1 Direct redox reactions in the preparation of bimetallic catalysts

In this case, metal-metal interactions result from surface redox reactions between the chemically prereduced metal « parent » M and the oxidized form of the modifier $M'^{Z'+}$.

According to Nernst's law, the equilibrium potential of the reversible oxidation of M :

$$M \Leftrightarrow M^{Z+} + ze$$

is

$$E_{M^{Z+}/M} = E^{\circ}_{M^{Z+}/M} + \frac{RT}{zF} \text{Ln} \frac{a_{M^{Z+}}}{a_M} \tag{1}$$

where $a_{M^{z+}}$ and a_M are the activities of oxidized (M^{Z+}) and reduced (M) species respectively and the other symbols have their usual meanings.

The equilibrium potential of the reversible reduction of $M'^{z'+}$

$$M'^{z'+} + z'e \Leftrightarrow M'$$

is given by

$$E_{M'^{z'+}/M'} = E^{\circ}_{M'^{z'+}/M'} + \frac{RT}{z'F} \text{Ln} \frac{a_{M'^{z'+}}}{a_{M'}}. \tag{2}$$

If the value of the free energy difference $\Delta(\Delta G)$ is negative $(\Delta(\Delta G) = - zz'F (E_{M'^{z'}/M'} - E_{M^{z+}/M}))$(3), $M'^{z'+}$ will be reduced by the metal M as follows :

$$z'M + z M'^{z'+} \rightarrow zM' + z' M^{z+}.$$

The system will tend to an equilibrium that corresponds to the equality of the two reversible potentials $E_{M'^{z'}/M'}$ and $E_{M^{z+}/M}$ The rate of redox reactions depends on the topography of the metallic crystal.

An example of the use of direct redox reactions in the preparation of bimetallic catalysts is the deposition of ruthenium on Raney copper [2, 3] according to the reaction :

$$3 \text{ Cu (s)} + 2 \text{ Ru}^{3+} \rightarrow 3 \text{ Cu}^{2+} + 2 \text{ Ru (s)}.$$

The modification of copper by redox reactions has been extended to other noble metals as Ag, Au, Pd, Pt, Ir, and Rh [2] and to metals supported on oxides such as alumina and silica.

1.2 Redox reactions of adsorbed species in the preparation of bimetallic catalysts (« Recharge method ») [4-14]

This method is so called by Szabò and co-workers « adsorption of metallic ions via ionization of adsorbed hydrogen »[4-7].

In this method, the ions of the modifier $M'^{z'+}$ are reduced by a reagent (most commonly hydrogen) that preadsorbs selectively on the metal.

For instance, in the case of metals which adsorb hydrogen, as platinum, the adsorbed hydrogen can be considered as a source of electrons :

$$H_{ads} \rightarrow H^+ + e.$$

Then the reduction of the modifier $M'^{z'+}$ can occur :

$$M'^{z'+} + z' H_{ads} \rightarrow M'_{ads} + z'H^+.$$

In this method, the prereduced monometallic catalyst is suspended in an electrolyte and treated with hydrogen until the surface of the catalyst is completely saturated by adsorbed hydrogen. Then, the reactor is flushed with inert gas to remove the hydrogen dissolved in the electrolyte. After outgassing, a deoxygenated solution containing the modifier $M'^{z'+}$ is added.

An example of the use of the « recharge » method is the preparation of a monodispersed Pt/Al_2O_3 by deposition of platinum onto a parent Pt/Al_2O_3 [8], of platinum-gold, platinum-palladium, platinum-rhenium bimetallic catalysts [12-14].

1.3 Catalytic reduction in the preparation of bimetallic catalysts

From a thermodynamic stand point, molecular hydrogen in solution can reduce any metallic salt that has a redox potential greater than that of the H_2/H^+ couple. However, at room temperature, the reduction of several metallic salts is kinetically limited.

The catalytic properties of the metal are used to increase the rate of reduction of the additive ion by means of a reducing agent, such as hydrogen, in solution. Thus the metal plays the role of the catalyst in the redox reaction :

$$M'^{z'+} + H_2 \xrightarrow{\text{catalyst}} M' + zH^+.$$

In such conditions, the additive is deposited on the catalytic site that is active for the reduction reaction. If the parent catalyst consists of a metal supported on a conductor, the additive will deposit competitively on the parent metal and on the support.

To date, the catalytic reduction of copper on platinum, on rhodium, on ruthenium and of rhenium on platinum has been investigated [14-16].

1. 4 Underpotential deposition [17, 18]

Underpotential deposition consists in the formation of a metal monolayer at potentials more positive than the reversible Nernst potential, that is before bulk deposition can occur.

In the bulk deposition of a metal M', according to the reaction :

$$M'^{z'^+} + z'e \Leftrightarrow M'.$$

the activity of the deposited metal is assumed to be constant and equal to one. In this case, the equilibrium potential is :

$$E_{M'^{z'+}/M'} = E^\circ_{M'^{z'+}/M'} + \frac{RT}{z'F} \, Ln \, a_{M'^{z'+}} \qquad (4)$$

When M' is deposited in submonolayer (M'_{ML}) on the metal M, the coverage θ of M by M' and then the activity of M' in submonolayer $a_{M'_{ML}}$ are less than one

$$a_{M'_{ML}} = \gamma_{M'} \, \theta \qquad (5)$$

where $\gamma_{M'}$ is the activity coefficient of M'.

For the deposition of a submonolayer of metal, the equilibrium potential can be given by :

$$E_{M'^{z'+}/M'} (\theta) = E^\circ_{M'^{z'+}/M'} + \frac{RT}{z'F} \, Ln \, \frac{C_{M'^{z'+}} \, \gamma_{M^{z'+}}}{\theta \, \gamma_{M'}}. \qquad (6)$$

The equilibrium potential E (θ) of the submonolayer is always more positive than the Nernst potential of bulk deposition. As a result, an underpotential deposition (UPD) of adatoms of M' on M can occur.

Underpotential deposition leads to formation of submonolayers before three dimensional bulk deposition occurs and the coverage varies with the potential and time of deposition. UPD is characterized by the existence of adatoms.

The formation of submonolayers of different adatoms by UPD differs substantially depending on the substrate structure and the crystal orientation. This finding shows up the large interest of this technique. Indeed with heterogeneous polycrystalline surface, the deposition potential can be controlled such that adatoms adsorb preferentially on specific sites of the polycrystal.

The simplest methods for investigating UPD are electrochemical.

Examples of the use of UPD technique are the modification of platinum catalyst by copper [10, 19], arsenic [20], gold [21], iron [22], lead and tin [23,24] and the modification of palladium [25, 26].

1.5 Concluding remarks

Surface redox reactions represent a relatively new approach to prepare supported bimetallic catalysts. Three main techniques have already been identified :
      - direct redox reactions between the parent metal and the modifier,
      - redox reactions between a reductant and the modifier (the reductant can be previously adsorbed on the parent metal or introduced as a reagent in the reactor (if no homogeneous reaction can occur between the modifier and the reductant)),
      - underpotential deposition.

All these variants are complementary to one another and by using different experimental conditions, different ligands, different reductants and different modes of preparation, this approach can apparently be used to prepare any type of bimetallic couple. But the catalytic properties of the final catalyst will depend strongly on the technique used. The selection of one variant over another depends on the desired structure of the bimetallic phase.

In conclusion, redox reactions allow the surface to be tailored during catalyst preparation, which explains the exceptional properties, particularly in terms of selectivity, of bimetallic catalysts prepared by these techniques.

2. CHARACTERIZATION OF METALLIC CATALYSTS BY THE MEANS OF ELECTROCHEMICAL METHODS

Transient electrochemical methods allow, through the measurement of the quantity of electricity obtained in response to a potential variation, to detect the surface modifications of the electrode due to the electrochemical formation or destruction of superficial layers.

One of the most widely used transient electrochemical methods is cyclic voltammetry with linear potential sweep. It consists in applying to the working electrode a potential E that varies in a linear way with time and recording the corresponding $i = f(E)$ curves.

A voltammogram is characteristic of the nature of the metal and of the adsorbed species in given experimental conditions.

So this method presents the advantage of characterizing the catalyst « in situ » and allows [10, 27-30] :
      - the determination of the initial area of the catalyst,
      - the determination of the quantity of adsorbed compound on the surface
         of the catalyst,
      - the determination of the adsorption stoechiometry,
      - the determination of the adsorption isotherms and thermodynamic
         parameters.

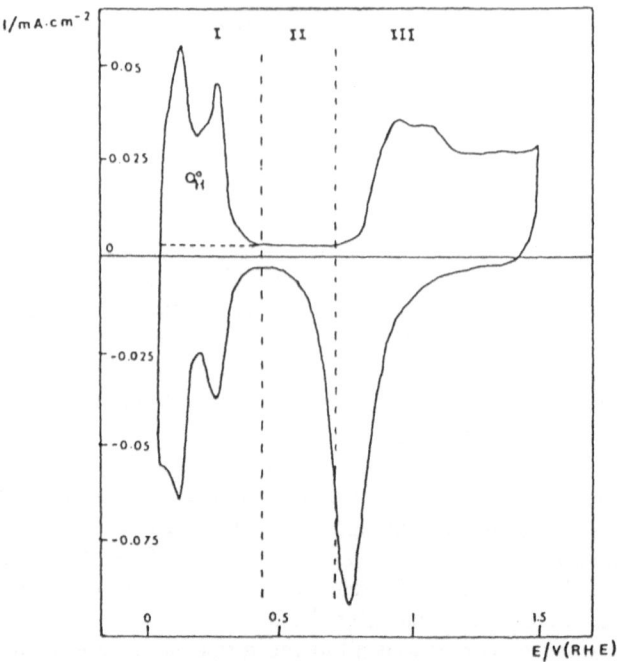

Figure 1 : Cyclic voltammogram of a platinized platinum catalyst electrode in 0.5 M H_2SO_4.
Sweep rate 50 mV.s^{-1}.[31]

The voltammogram of platinum can be divided into three regions (Fig. 1) :
- **region I** corresponding to the adsorption (i < 0 : $H^+ + e^- \rightarrow H_{ads}$) and the desorption   (i > 0   : $H_{ads} \rightarrow H^+ + e^-$) of hydrogen (from 0.05 to 0.5 V/RHE). The perfect symmetry of the anodic   and   cathodic curves confirms the reversibility of the equilibrium $H^+ + e^- \Leftrightarrow H_{ads}+$ on platinum.
- **region II** in which no electrochemical reaction occurs (from 0.5 to 0.65 V/RHE).
- **region III**   corresponding   to   the   adsorption   (i > 0 : $H_2O \rightarrow O_{ads} + 2 H^+ + 2e^-$) and desorption (i < 0 : $O_{ads} + 2 H^+ + 2e^- \rightarrow H_2O$) of oxygen (from 0.65 to 1.5 V/RHE). The lack of reversibility of the adsorption and desorption peaks indicates that the kinetics of these two opposites are slow.

2.1 Determination of metallic surface area by cyclic voltammetry

Through integration of the anodic or cathodic part relative to the hydrogen potential region of the voltammogram [28], it is possible to obtain the quantity of electricity, Q_H^O, associated with an hydrogen monolayer and to calculate the initial number, N_o, of the surface platinum atoms (see Fig. 1) :

$$N_o = \frac{Q_H^O \ (\mu C)}{1.6.10^{13}} \tag{7}$$

or the initial surface area (assuming that a platinum atom occupies an average area close to 7.6 $\mathring{A}^2$, i.e $1.31.10^{15}$ atoms per cm²).

2.2 Thermodynamic study of the adsorption of different compounds by cyclic voltammetry

The free energy of adsorption ΔG can be determined by means of transient electrochemical methods. As a matter of fact, the electrooxidation or the electroreduction of the adsorbed species is a function of the free energy of adsorption given by Gibbs equation :

$$\Delta G = -zFE \tag{8.}$$

The partial derivative of free energy in relation to the temperature, in a reduced range of temperature, can be written :

$$\frac{\partial \Delta G}{\partial T} \ = \ -zF \ \frac{\partial E}{\partial T} \ = \ - \Delta S \tag{9}$$

The ratio $\dfrac{\partial E}{\partial T}$ can be determined by making the measurement of two values of the potential of the catalyst, at two different temperatures.

Consequently the energy of adsorption ΔH can be deduced from ΔG and ΔS measurements.

2.2.1 Thermodynamic study of hydrogen adsorption on platinum

Reactions of electroreduction of protons or electrooxydation of hydrogen are fast on platinum. As a consequence the equilibrium corresponding to the overall reaction :

$$H_2 \Leftrightarrow 2H^+ + 2e^-$$

is assured at every moment.

Thus, any catalyst working in liquid phase and in the presence of hydrogen behaves like a reversible hydrogen electrode whose potential obeys Nernst equation

$$E = 0, 06 \log [H^+] - 0, 03 \log pH_2 \tag{10}$$

Hence, the measurement of the potential is a measurement of the hydrogen pressure near the catalyst and therefore of hydrogen coverage. The cyclic voltammetry curves (see Fig. 1) allow, for any potential value, to define the amount of hydrogen adsorbed at that same potential (by integration of the amount of electricity associated with the reaction $H^+ + e^- \rightarrow H_{ads}$).

The potential at which the electrochemical reaction occurs can be correlated to the free energy corresponding to this reaction and given by the Gibbs equation :

$$\Delta G = - zFE = \Delta H - T\Delta S \tag{11}$$

The voltammogram (see Fig. 1) shows two peaks of hydrogen adsorption or desorption corresponding to two different potentials, hence to different energies of adsorption. The first peak, relative to a strongly bound hydrogen, appears at a potential $E = 0.26$ V/RHE (low hydrogen coverage). The second peak which corresponds to a weakly bound hydrogen appears at 0.13 V/RHE (high hydrogen coverage).

The above potentials are measured in relation to a reference electrode, so it is not possible to define the true energy of adsorption corresponding to the different forms of adsorbed hydrogen.

However, the potential difference between the two peaks (0.13 V) shows that there is, between each form of hydrogen, a difference of energy of adsorption, equal to :

$$\begin{aligned}
\Delta (\Delta G) &= zF \, \Delta E \tag{12} \\
&= 2 \times 96\,500 \times 0.13 \\
&\cong 25 \text{ kJ/mol.}
\end{aligned}$$

On the other hand the temperature coefficient $\dfrac{\partial E}{\partial T}$ allows to calculate the entropy variation related to the hydrogen adsorption reaction. Table I gives the temperature coefficients and the entropy variations relative to hydrogen both strongly and weakly bound.

Table I : Temperature coefficients and entropy variations relative to hydrogen adsorption on platinum[32].

Hydrogen	$\dfrac{\partial E}{\partial T}$ (V.K^{-1})	ΔS (J.K^{-1}mol^{-1})
weakly bound	2.74. 10^{-3}	- 530
strongly bound	3.21. 10^{-3}	- 620

It must be remarked that the entropy variations are negative, which correspond to a loss of degrees of freedom when the hydrogen molecule in gas phase passes into the state of adsorbed hydrogen atoms. Moreover the weakly bound hydrogen has lost a smaller degree of freedom during its adsorption and so it is more mobile on the surface of the catalyst. Moreover

$$\Delta (\Delta H) = \Delta (\Delta G) + T\Delta (\Delta S) \cong 50 \text{ kJ/mol} \qquad (13)$$

This value is comparable to the one given in the literature [33, 34].

The cyclic voltammetry curves (see Fig. 1) allow, for any potential value, to define the amount of hydrogen adsorbed. On the other hand, according to Nernst equation for a given pH, (equation 10), the potential is directly linked to the hydrogen pressure.

Consequently, it is possible to plot the hydrogen adsorption isotherms (see Fig. 2).

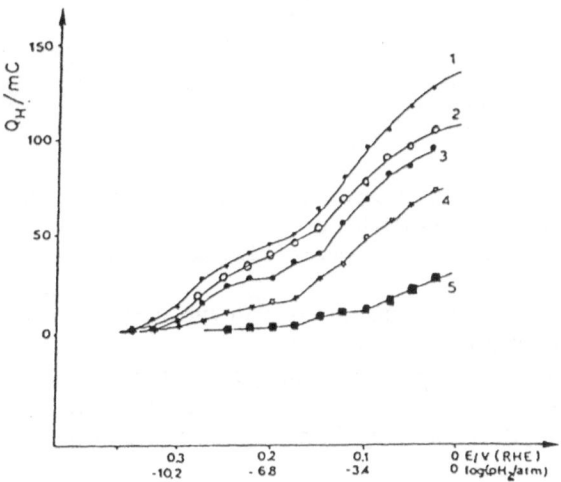

Figure 2: Adsorption isotherms of hydrogen on platinum modified by sulphur at different sulphur coverages [29]   (1) 0; (2) 0.04, (3) 0.07, (4) 0.17, (5) 0.73.

The isotherm (see Fig.2) is formed by two waves which are in relation with the two forms of hydrogen previously described. Each of these waves follows a Temkin type law which allows the determination of the free energy of hydrogen adsorption on the catalyst.

2.2.2 Thermodynamic study of sulphur adsorption on platinum

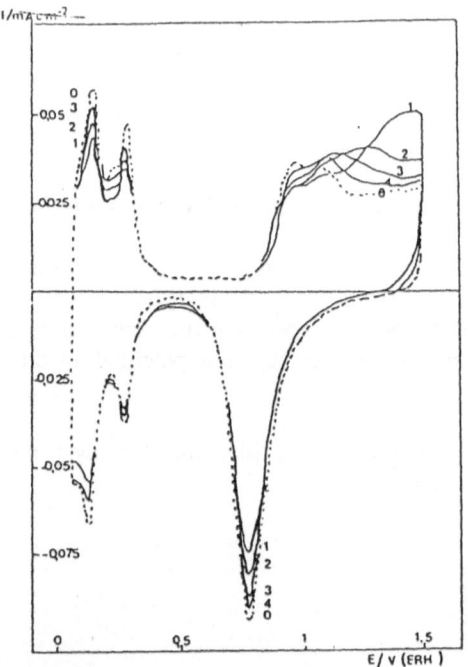

Figure 3 : Cyclic voltammogram curves of platinized platinum electrode modified by sulphur during successive cycles (H₂SO₄ 0.5 M; 50 mV/s; 25°C; θs = 0.27) [30], —— in absence of adsorbed sulphur, ___ in presence of adsorbed sulphur.

Sulphur adsorbed on a platinized platinum electrode modifies the cyclic voltammetry curves characteristic of this metal.

In the adsorption region of hydrogen, a decrease of the high of the adsorption or desorption peaks shows that a fraction of the surface has been occupied by sulphur, thus inhibiting the adsorption of hydrogen, (Fig. 3).

Moreover, the sulphur coverage can be introduced by the relation :

$$\theta_S = \frac{Q_H^O - Q_H^S}{Q_H^O} \qquad (14)$$

where Q_H^0 and Q_H^S are the quantities of electricity associated with the oxidation of adsorbed hydrogen in the absence and in the presence of sulphur.

The quantity of adsorbed sulphur can be determined by its electrooxidation which leads to its desorption to the state of sulphate [35-36]. This electrochemical study allows to determine the oxidation state and the stoichiometry of adsorbed sulphur on platinium [30, 31].

The effect of preadsorbed sulphur on the adsorption of hydrogen on platinum allows to obtain the evolution of the free energy of adsorption of hydrogen on platinum (for both strongly and weakly bound hydrogen) as a function of sulphur coverage, defined by the isotherms of hydrogen adsorption (Tab.II).

Table II : Effect of sulphur on the free energy of adsorption of hydrogen on platinum [29]

N_S/N_{Pt}	Strongly bound hydrogen - $\Delta G°$ (kJ/mol)	Weakly bound hydrogen - $\Delta G°$ (kJ/mol)
0	37	14
0.04	36	13
0.07	31	12
0.17	30	7
0.72	38	0.5

Sulphur doesn't modify the free energy of adsorption of strongly bound hydrogen. But on the other hand, it affects strongly the adsorption equilibrium constants of weakly bound hydrogen.

2.2.3 Thermodynamic study of maleic acid adsorption

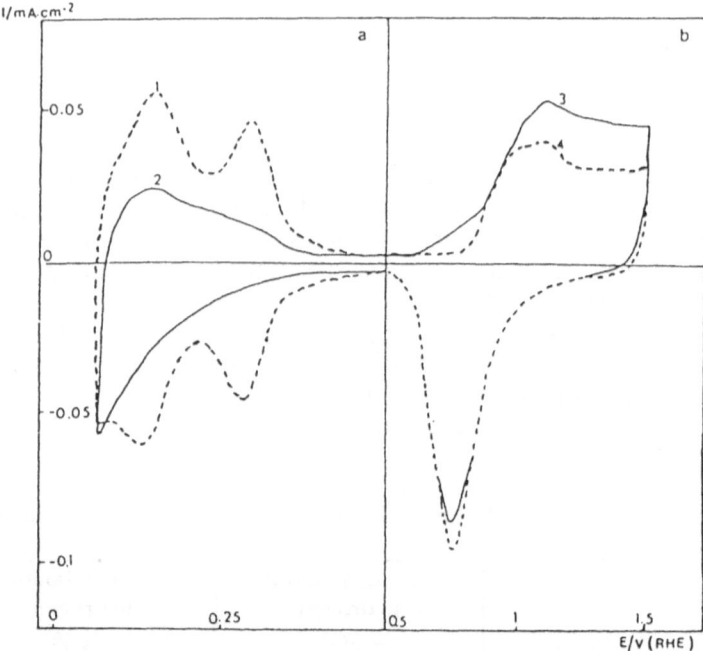

Figure 4 : Effect of the adsorption of maleic acid on the cyclic voltammogram of platinized platinum catalyst electrode (0.5 M H$_2$SO$_4$; 50 mV/sec) [31] (——)in absence of adsorbed maleic acid, (____) in presence of adsorbed maleic acid; (a) cyclic voltammogram in the region of adsorption- desorption of hydrogen;(b) cyclic voltammogram in the region of adsorption- desorption of oxygen.

Figure 4a shows a voltammogram recorded in the region of hydrogen after adsorption of maleic acid. The decrease of the hydrogen peaks proves that hydrogen and unsaturated compounds adsorb in competition on the same sites.

A comparison of the two curves, 3 and 4, of Figure 4, recorded between 0.5 and 1.5 V/RHE, in the presence and the absence of adsorbed maleic acid, shows that the oxidation of this adsorbed compound causes an increase of the anodic current.

Through integration of the difference between curves 1 and 2, the degree of coverage, θ_M, of the catalyst by maleic acid can be determined as the fraction of the surface deactivated for hydrogen chemisorption.

$$\theta_M = \frac{\left(Q_H^O - Q_H^M\right)}{Q_H^O} \tag{17}$$

where Q_H^O and Q_H^M are the quantities of electricity associated with the oxidation of adsorbed hydrogen in the absence and in the presence of maleic acid.

Furthermore, adsorption isotherms of maleic acid can be determined by measurement of θ_M for different acid concentrations (Fig. 5).

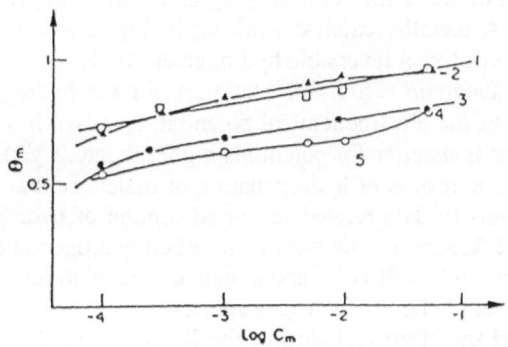

Figure 5 : Adsorptions of maleic acid ($C= 10^{-2}$ M) on Pt/Pt electrodes of different roughness parameters : (1) 2.3,(2) 22, (3) 58, (4) 300, (5) 360 [38].

The quantity of maleic acid adsorbed on a platinum electrode can be determined by integration of the difference between curves 3 and 4 of Figure 4, which leads to the quantity of electricity Q_{ox}^M associated to the oxidation of this adsorbate, according to reaction :

$$
\begin{array}{cc}
\text{HOOC} & \text{COOH} \\
\quad C = C \quad & + 4\,H_2O \quad \rightarrow \quad 4CO_2 + 12H^+ + 12e^- \\
\text{H} & \text{H}
\end{array}
$$

If n_{Pt_M} is the number of platinum sites occupied by one molecule of adsorbed maleic acid, it can be written :

$$\frac{Q_H^O - Q_H^M}{n_{Pt_M}} = \frac{Q_{ox}^M}{12} \tag{16}$$

It has been found [10, 31] that one molecule of adsorbed maleic acid occupies about 5 accessible platinum atoms.

2.3 Electrochemical potential of the catalyst during hydrogenation reactions in liquid phase

Figure 6 and 7 [37] show the evolution of the potential of platinum catalyst and of maleic acid concentration versus working time during an hydrogenation reaction of maleic acid. It can be noted a sudden increase of the potential during the maleic acid introduction. This variation corresponds to a diminution of hydrogen coverage on the catalyst. The more the reaction progresses the more the potential tends to its initial value. A metallic catalyst working in liquid phase and in the presence of hydrogen behaves like a reversible hydrogen electrode.

The change of the catalytic activity of platinum, for the hydrogenation of maleic acid, as a function of the electrochemical potential, is shown in Figure 8. It can be seen that the catalyst is inactive for potentials higher than 0.2 V/R.H.E..

The comparison of the results of hydrogenation of maleic acid at controlled potentials with the voltammetry data related to the adsorption of hydrogen shows that at potentials below 0.2 V/R.H.E. only weakly adsorbed hydrogen is consumed, whereas at potentials above 0.2 V/R.H.E. about half of the platinum surface is covered by very inactive strongly bound hydrogen species.

It can be concluded therefore that during the liquid- phase hydrogenation reactions of C=C bonds, only weakly bound hydrogen species are active. Moreover this hydrogen is also most mobile on the platinum surface.

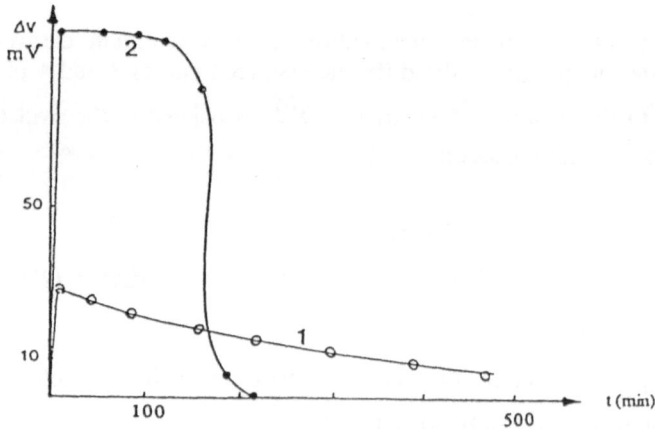

Figure 6 : Evolution of the potential of a Pt catalyst versus working time during hydrogenation reaction [37] (Curve 1 : 25 mg of Pt, curve 2 : 100 mg of Pt).

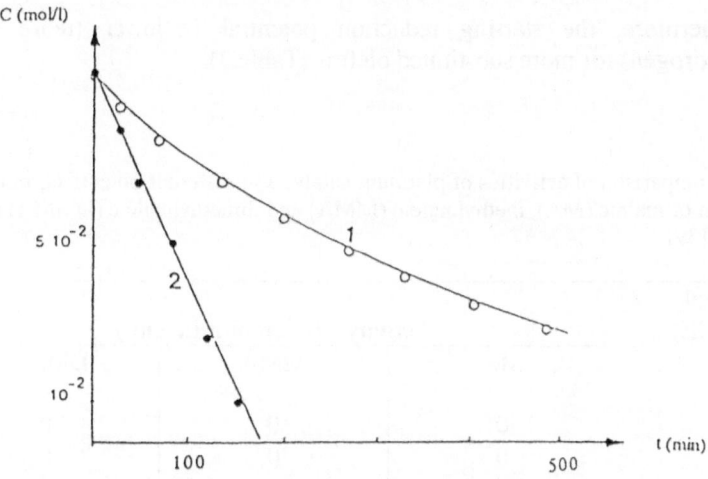

Figure 7 : Evolution of maleic acid concentration versus working time during hydrogenation reaction [37] (Curve 1 : 25 mg of Pt, curve 2 : 100 mg of Pt).

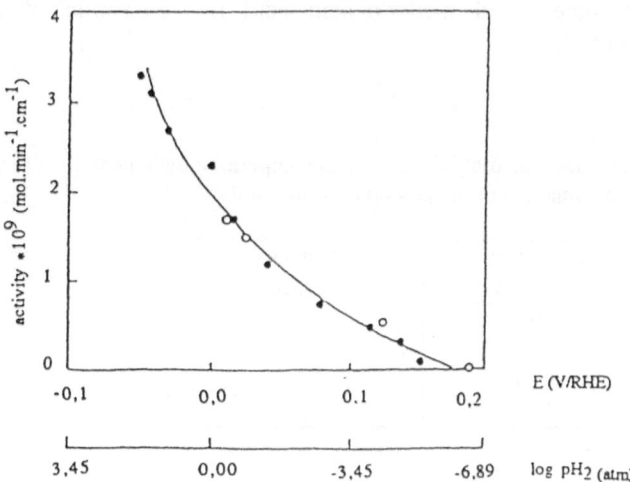

Figure 8 : Catalytic activity of platinum for the hydrogenation of maleic acid as a function of the potential of the catalyst [38]

Furthermore, the starting reduction potential is lower (more weakly adsorbed hydrogen) for more substituted olefins (Table 3)

Table III : Comparison of activities of platinum catalysts at different potentials, in catalytic hydrogenation of maleic (MA), methylmaleic (MMA) and dimethylmaleic (DMMA) acids ($C = 10^{-3}$ M) [39].

Potential (mV/RHE)	Activity . 10^9 (mol/min. cm²)		
	MA	MMA	DMMA
254	0	0	0
204	0	0	0
184	0.06	0	0
154	0.09	0.07	0
124	0.21	0.16	0.06
103	0.23	0.19	0.16
93	0.29	0.22	0.19

Thanks to which it is possible to control the selectivity, in competitive hydrogenation of maleic and methylmaleic acid by imposition of different potentials (Tab. IV) [39].

Table IV :Variation of the selectivity S_{MA}/S_{MMA} in competitive hydrogenation of maleic and methylmaleic acids as a function of the potential of the catalyst.

Potential (mV/RHE)	Selectivity S_{MA}/S_{MMA}
9	1.7
69	3.1
150	7.0
171	50

On the other hand, Table V shows the change of the selectivity as a function of the potential of the catalyst in the hydrogenation of acetylenic compounds such as butynedioic acid on platinum. It is worth noting that weakly bound hydrogen which is active in olefinic compounds hydrogenation leads to a low selectivity in alkene during the hydrogenation of acetylenic compounds [32].

Table V : Variation of the selectivity in olefinic compounds during hydrogenation of butynedioic acid as a function of the potential of the catalyst [32].

Potential	Selectivity in olefinic compounds C = C/ C - C %
9	6
4	7
86	26

Moreover, the strongly bound hydrogen, is more active for the hydrogenation of acetylenic bonds, than for the hydrogenation of the double bound. This allows to explain the increase in the selectivity in alkene as soon as the coverage in weakly bound hydrogen decreases.

2.4 Concluding remarks

The characterization of the surface state and of the chemisorbed species on metallic catalysts working in liquid phase can be carried out « in situ » by a transient electrochemical method : the linear potential sweep cyclic voltammetry. The use of that technique allows :
- the determination at the initial metallic area of the catalyst,
- the measurement of the surface area occupied by reagents or by modifiers,
- the evaluation of the adsorption stoechiometry,
- the measurement of the equilibrium coverage of the adsorbed species which allows to obtain the adsorption isotherms and thermodynamics of that adsorption.

On the other hand, during catalytic hydrogenation of unsaturated compounds, the activity and selectivity of platinum catalysts depends on the value of working potential of the catalyst.

Using such electrochemical studies of catalytic hydrogenation reactions pointed out that only weakly adsorbed hydrogen species are involved in hydrogenation of olefinic compounds on platinum catalysts.

Acknowledgements
The authors are gratefull to all the persons of the laboratory of whom the work allows us to write this course :
    Permanent Researchers : E. Lamy-Pitara, P. Marecot, J.C. Menezo, C. Montassier.
    Students : L. Bencharif, L. El Ouazzani-Benhima, M.E. Gonzalez, L. Lghouzouani, S. Moukolo, J. Naja, S. Peyrovi, C.L. Pieck, Y. Tainon.

References

[1]   J. Barbier, Redox Reactions in the Tailoring of Bimetallic Catalysts.
Advances in Catalysts Preparation, Catalytica Studies Division (1992) 3.
[2]   C. Montassier, J.C. Ménézo, S. Moukolo, J. Naja, J. Barbier. Heterogeneous
Catalysis and Fine Chemicals II, Stud. Sci. Catal., M. Guisnet, J. Barrault. C.
Bouchoule, D. Duprez, G. Perot, R. Maurel, C. Montassier Eds. **59**, (1991) 223.
[3]   J. Barbier, J.P. Boitiaux, P. Chaumette, S. Leporq, J.C. Ménézo. C.
Montassier, EP 380, 402, assigned to Institut Français du Petrole (1990).
[4]   S. Szabó, F. Nagy, J. Electroanal. Chem., **84**, (1977) 93.
[5]   S. Szabó, F. Nagy, J. Electroanal. Chem., **85**, (1977) 339.
[6]   S. Szabó, F. Nagy, J. Electroanal. Chem., **87**, (1978) 261.
[7]   S. Szabó, F. Nagy, J. Electroanal. Chem., **160**, (1984) 299.
[8]   J.C. Ménézo, M.F. Denanot, S. Peyrovi, J. Barbier. Appl. Catal.. **15**,(1985)
353.
[9]   S. Szabó, F. Nagy, React. Kinet. Catal. Lett., **351** (2), (1987)133.
[10]   E. Lamy-Pitara, J. Barbier, C. Lamy, J. Chim. Phys., **77** (10) (1980) 967.
[11]   J. Margitfalvi, S. Szabó, F. Nagy, Stud. Surf. Sci. Cat.. **17** (1986) 373.
[12]   S. Szabó, F. Nagy, Isr. J. Chem., **18**, (1979) 162.
[13]   J. Margitfalvi, S. Szabó, F. Nagy, S. Gobolos, M. Hegedus, Preparation of
catalysts III, Stud. Surf. Sci. Catal., G. Poncelet, P. Grange. P.A. Jacobs,
Eds, Elsevier, Amsterdam, **16**, (1983).
[14]   C.L. Pieck, P. Marecot, J. Barbier, Appl. Cat. A, **134**, (1996) 319.
[15]   C.L. Pieck, P. Marecot, C.A. Querini, J.M. Parera, and J. Barbier, Appl.
Cat. A, **133**, (1995) 281.
[16]   C.L. Pieck, Ph. D. Thesis, University of Poitiers (1994).
[17]   D.M. Kolb, Adv. Electrochem. Electrochem. Engineer., **11** (1978)125.
[18]   S. Szabó, International Reviews in Physical Chemistry. **10** (2). (1991) 207.
[19]   S.H. Calde, S. Buckenstein, Anal. Chem., **43**, (1971)1858.
[20]   N. Furuya, S. Motoo, J. Electroanal. Chem., **72**, (1976)165.
[21]   N. Furuya, S. Motoo, J. Electroanal. Chem., **78**, (1977) 243.
[22]   N. Furuya, S. Motoo, J. Electroanal. Chem., **88**, (1978)151.
[23]   E. Lamy-Pitara, L. El Quazzani-Benhima. and J. Barbier, J. Electroanal.
Chem., **335**, (1992) 363.
[24]   N. Furuya, S. Motoo, J. Electroanal. Chem., **98**, (1979) 195.
[25]   I. Bakos, S. Szabo, F. Nagy. T. Mallat, Z. Bodnar, J. Electroanal. Chem.,
309, (1991) 203
[26]   I. Bakos, S. Szabo, React. Kinet. Catal. Lett., **41**, (1990) 53.
(1976) 89.
[27]   E. Lamy-Pitara and J. Barbier, Electrochimica Acta, **27** (6) (1982) 713.
[28]   R. Woods, Chemisorption at electrodes : hydrogen and oxygen on noble
metals and their alloys, Electroanal. Chem., Vol.9, ed. A.J. Bard, Marcel
Dekker Inc, N.Y. (1976)
[29]   E. Lamy-Pitara, L. Lghouzouani, Y. Tainon, J. Barbier. J. Electroanal.
Chem, 260, 157 (1989)

[30]   E. Lamy- Pitara, L. Bencarif, J. Barbier, Electrochem. Acta, **30**, 971 (1985)

[31]   E. Lamy- Pitara, L. Bencharif, J. Barbier, Appl. Cat., **18**, 117 (1985)

[32]   J. Barbier, E. Lamy- Pitara, P. Marecot, Bull. Soc. Chim. Bel., **105**, 99 (1996).

[33]   K.R. Cristmann in « Hydrogen Effects in Catalysis ». ed. Z. Paal and P.G. Menon, M. Dekker, N.Y. 3 (1988).

[34]   K. Cristmann and G. Ertl Surf. Sci. **60**, (1976) 365.

[35]   A.A. Sutyagina, B.A. Perepelitsa and M.N. Semenko, Rh. Fiz. Khim., **57** (3) , (1985). 681

[36] A.A. Sutyagina, B.A. Perepelitsa and M.N. Semenko, Rh. Fiz. Khim., **57** (3) (1985) 685

[37]   J. Barbier, M. E Gonzalez, J. Chim. phys, **76** (1979)

[38]   E. lamy-Pitara, I. Belegridi, L. El Ouazzani-Benhima, J. Barbier, Catalysis Letters, **19**, (1993) 87.

[39]   E. Lamy-Pitara, I. Belegridi, J. Barbier, Catalysis Today, **24**, (1995) 151.

Catalysis and Automotive Pollution Control

J. Barbier, C. Micheaud, E. Rohart and E. Lafitte

LACCO, Unité de Recherche Associée au CNRS, DO 350, Université de Poitiers, 40 avenue du Recteur Pineau, 86022 Poitiers cedex, France

---

1. INTRODUCTION

Beginning in the mid-70s, it has become a general concern to reduce the level of exhaust emissions from motor vehicles. In the early days of exhaust gas catalysts, the principle requirement was to oxidise unburnt hydrocarbons (HC) and carbon monoxide (CO). Later legislation has become stricter and has focussed on additionnaly removing nitrogen oxides (NO_x) [1-3].

The fitting of platinum group metals catalytic converters appeared to be the only efficient tool to achieve the standard emissions levels.

$$HC \rightarrow H_2O + CO_2 \qquad (1)$$

$$CO \rightarrow CO_2 \qquad (2)$$

$$NO_x \rightarrow N_2 + H_2O \qquad (3)$$

At the beginning of 21th century, legislation on air environment and essentially on automotive pollution control will be stricter. This paper deals with the catalytic reactions and the solid catalysts involved in such exhaust gas converters.

In the first chapter, we will discuss on general aspects of catalytic oxidation of CO, HC and oxygenated compounds.

In the second one, we will focus our attention on the role of water in the combustion of CO and unburnt hydrocarbons.

In the third one, we will deal with the reduction of NO_x in stoichiometric, rich and lean conditions.

2. GENERAL ASPECTS

Catalytic oxydation of CO or HC on platinum group metals has received considerable attention over the past years.

Figure 1 shows the evolution of the catalytic activity for total oxidation and the nature of the oxide under consideration. It is interesting to note that the lower the heat of adsorption of oxygen, the better the activity in CO and HC combustion [4-6].

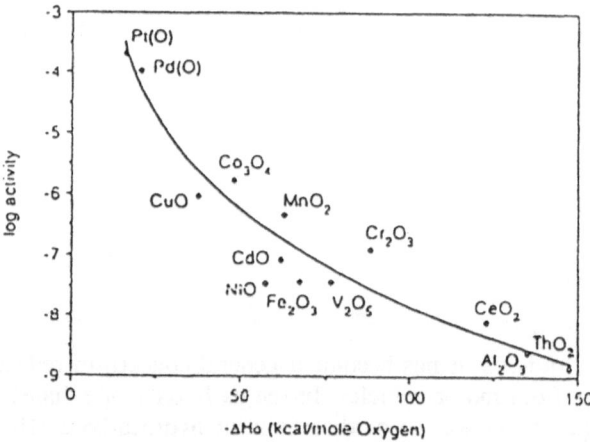

Figure 1 : Complete oxidation - Catalytic activity versus the heat of formation of the oxide

Furthermore, it is well known that catalytic oxidation is controlled by a <u>redox mechanism</u>. An electronic transfer (Scheme 1) between chemisorbed oxygen and metal or between oxygen and reactants has been traditionnaly suggested to describe the reaction mechanism (Scheme 2).

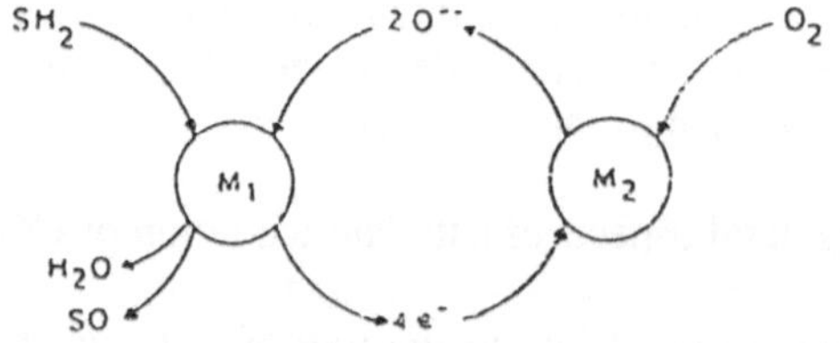

$$CH_4 + * \rightarrow \underset{*}{\overset{||}{CH_2}} + 2\ H^*$$

$$Pt = O \Rightarrow Pt,\ \overline{O}| + CH_2 \rightarrow \quad \underset{H}{\overset{H}{\underset{/}{\overset{\backslash}{C = O}}}}$$

Scheme 1 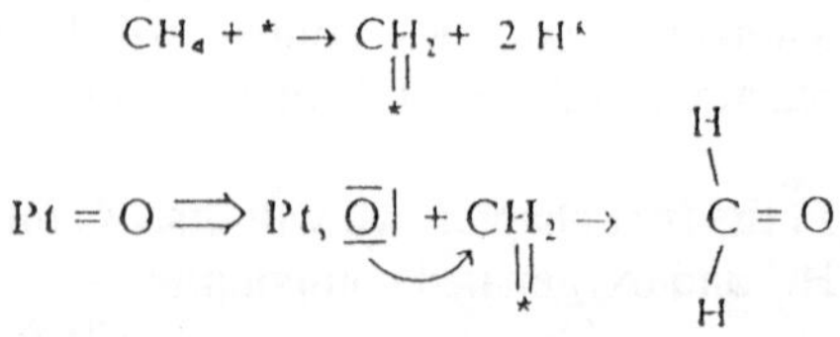   Scheme 2

3. CATALYTIC COMBUSTION

3. 1 Catalytic oxidation of carbon monoxide

Concerning the oxidation of CO, the CO_2 formation at low temperature can occur either by involving the surface reaction of one adsorbed reactant with the other one in gas phase (mechanism of Eley-Rideal, equations 2 and 3), or the reaction of CO and O together in adsorbed state on the catalyst (mechanism of Langmuir-Hinshelwood, equation 1).

$$O_{ads} + CO_{ads} \rightarrow CO_2 \qquad (1)$$

$$O_{ads} + CO_{gas} \rightarrow CO_2 \qquad (2)$$

$$O_2 + 2CO_{ads} \rightarrow 2CO_2 \qquad (3)$$

A complete description of the reaction mechanism is only possible by understanding the different steps and in particular the adsorption of the reactants.

3.1.1 Adsorption of carbon monoxide

Adsorption normally takes place in molecular form. The energy of adsorption of CO on the different transient metals varies between 26 and 40 kcal/mol which is comparable to the binding energy of CO in carbonyls. On the other hand, that energy of adsorption is widely dependent on the CO coverage (Fig. 2) [7].

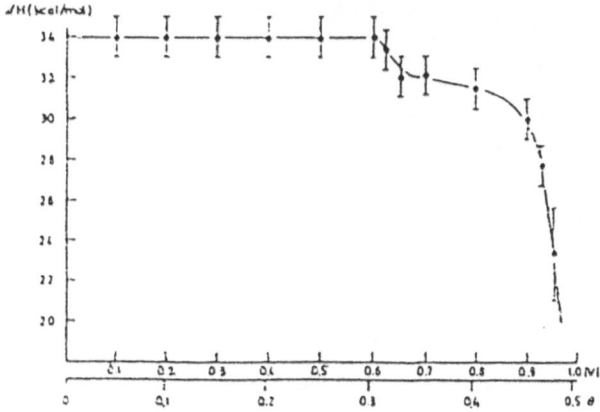

Figure 2 : Heat of adsorption of CO as the function of the coverage degree θ

At low coverage, less than 30%, heat of adsorption is constant but decreases drastically for $\theta = 0.5$, due to dipolar interactions between chemisorbed CO.

Investigations of the kinetics of an adsorption or a desorption process enable to obtain interesting informations about the state of the bonding in the adsorbed phase.

The rate of adsorption onto a solid surface is given by the Knüdsen equation :

$$r_a = p/(2\Pi mkT)^{1/2} \, S(\theta/\theta_{max})$$

where p is the gas pressure, m the adsorbate mass, k the Boltzmann constant, T the absolute temperature, S the sticking coefficient.

This definition is based on the assumption that the rate of desorption is negligible. The experimental approach of r_a is however only possible if :

        - coverage $\theta = 0$,

        - low temperature (<600K),

        - high partial pressure of CO,

        - $S(\theta/\theta_{max})$ close to the unity.

In these conditions, the adsorption is a nonactived process (Activation energy, $E_A=0$).

In addition : $E_{ads} = E_d - E_A$ so $E_{ads} = E_d = -\Delta H_{ads}$

where E_d is the desorption energy, E_{ads} the adsorption energy, and ΔH_{ads} the adsorption enthalpy.

Nevertheless, interesting information on the mechanism of adsorption dynamics is contained in the functionnal dependence of the sticking coefficient on coverage (Fig. 3). S decreases drastically for $\theta > 0.15$ [8].

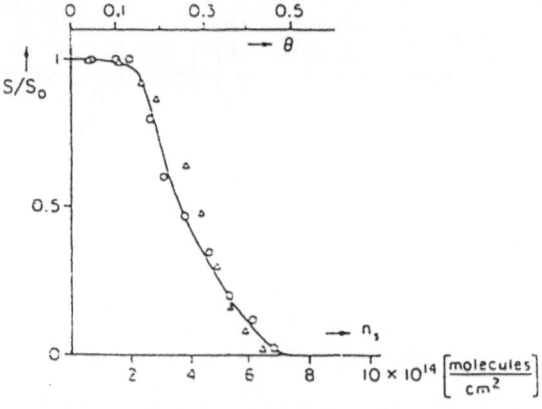

Figure 3 : The relative sticking coefficient as a function of the CO coverage on Pt(111)

For desorption, a more direct way of determining the kinetic parameters consists of measuring the mean residence time τ of adsorbed particles on the surface :

$$\tau = 1/r_d \qquad \text{where } r_d \text{ is the desorption rate.}$$

3.1.2 Adsorption of oxygen

Oxygen is dissociatively adsorbed on the platinum group metals at temperatures above 100 K, as may be concluded from isotopic exchange experimentations. The formation of bulk oxide often takes place under high temperatures conditions (more than 800 K).

A distinction may be made between chemisorbed oxygen and oxygen bound in the form of bulk oxides. One of the easiest method to distinguish between chemisorbed and incorporated oxygen is their reactivity towards H_2 or CO, incorporated oxygen is removed only at higher temperatures.

For example, the activation energy for CO oxidation differs from 45 kcal/mol between chemisorbed oxygen on iridium and iridium oxide [9]. From this result, there is no more doubt that chemisorbed oxygen is the active species in the catalytic oxidation of CO. Heat of adsorption ranges between 80 and 50 kcal/mol, whatever the metal may be. This means that the strength of the M-O bound is much higher than M-CO. Nevertheless, it is suggested that adsorbed oxygen atoms exhibit a much lower surface mobility than adsorbed CO.

The examination of the way in which the heat of adsorption depends on coverage, has revealed that the heat of desorption falls slowly with rising coverage (Fig. 4) [10]. Temperature at which thermal desorption occurs increases with rising oxidation temperature. suggesting once again the formation of bulk oxide layers (Fig. 5).

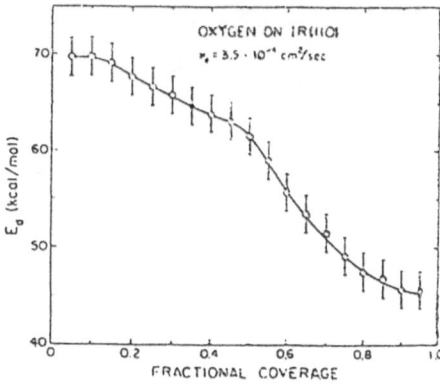

Figure 4 : Heat of desorption of O_2 versus coverage degree

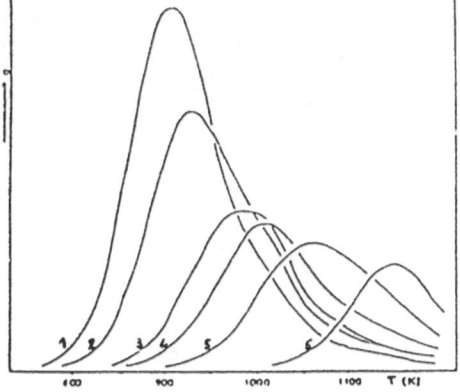

Figure 5 : Thermal desorption spectra for oxygen adsorption on Pd(111). The adsorption temperature increases in sequence 1-6

3.1.3 *Experimental approach of catalytic oxidation of CO*

We focus our attention on catalytic reactions involving Pd and Pt.

Theoretically, oxygen interacts stronglier with metal than CO, suggesting that no CO may be adsorbed at the equilibrium. In fact experimental results (Fig. 6) are at variance with the theory, CO inhibiting itself the adsorption of oxygen for high partial pressures [11-12].

This controversy may be explained by taking into account the larger mobility of CO on the surface of the metal than oxygen (see *3.1.2*).

From the observed results, it may be concluded that oxygen must be adsorbed in the atomic state on the surface to react (molecular oxygen desorbs with a broad maximum which is located at about 520°C then for a range of temperature higher than this concerned). The type of Eley-Rideal reaction, $O_2 + 2CO_{ads} \rightarrow 2CO_2$, can easily be ruled out. No CO_2 formation takes place if a surface saturated with adsorbed CO is exposed to gaseous O_2. Obviously, the dissociative chemisorption of oxygen is a necessary prerequisite.

The following discussion will be centered on the experimental distinction between the Langmuir-Hinshelwood and the Eley-Rideal types of mechanisms for product formation.

Figure 7 shows the rate of CO_2 production on a Pd(111) surface as a function of temperature for different ratios P_{CO}/P_{O_2}. It is seen that the reaction rate increases rapidly after which it slowly decreases with increasing temperature [13]. The temperature corresponding to the maximum rate increases with increasing P_{CO}/P_{O_2}.

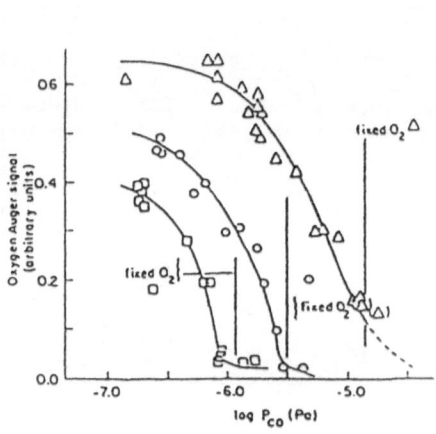

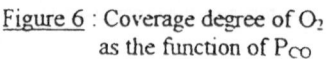

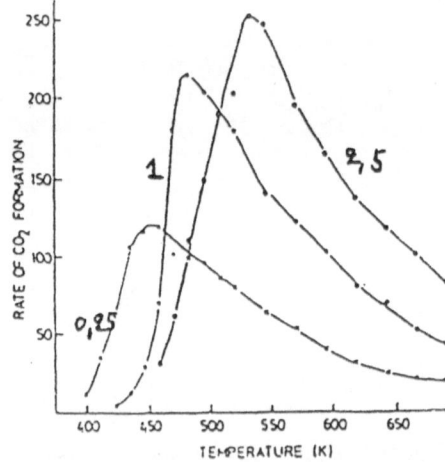

Figure 6 : Coverage degree of O_2 as the function of P_{CO}

Figure 7 : Rate of CO_2 formation as the function of the temperature for differents P_{CO}/P_{O_2}

The following conclusions may be drawn from figure 7 :

- The reaction rate (r) at first increases with the temperature which is in good agreement with an increase of rate constant according to the arrhenius equation.

- At the maximum of r (T $\approx$ 200°C), it may be suggested that the coverage of active species , such as oxygen, decreases (either by diffusion into the metal or by desorption). It could however be shown that these processes are negligible at T$\leq$700K. In fact, this loss of activity is in relation with the activation of CO.

In this range, the CO coverage under reaction conditions decreases continously due to progressive desorption and becomes practically zero at the maximum rate.

It may be assumed that the reaction is governed by a Langmuir-Hinshelwood mechanism.

Kinetic studies were reported by Ku and Bonzel [14] and may be summerized as follow :

Table I : Kinetic orders for the catalytic oxidation of CO versus temperature

Temperature range	O_2 order	CO order
T < Tmax	1	1 if $Pco<Po_2$ Negativ if $Pco>Po_2$
T > Tmax	0 if $Po_2 \gg Pco$ 1 if $Po_2 \ll Pco$	1

3.2 Catalytic oxidation of hydrocarbons (HC)

Noble metals such as Pt, Pd and Rh are the more active for HC oxidation.

3.2.1 Alkane oxidation

Alkane oxidability is strongly dependent of the nature of the metal. Thus Pd is more active than Pt for methane and ethane oxidation which is different for heavier compounds.

Hydrocarbons oxidation is a very structure sensitive reaction [15-17]. The lower the dispersion the better the activity (Tab. II).

Table II : Catalytic oxidation of methane : TON (s^{-1}) as the function
of metallic accessibility for Pd/Al_2O_3 and Pt/Al_2O_3 catalysts

Dispersion %	TON s^{-1}
100	0.004
70	0.013
33	0.150

Pd/Al_2O_3

Dispersion %	TON s^{-1}
90	0.001
6	0.020

Pt/Al_2O_3

From a kinetic point of view, it seems that the orders of the reactants (such as propane) are in relation with the nature of the metal. Several studies claim that on supported Pd and Rh catalysts, O_2 and HC compete to react on the surface. On platinum, O_2 interacts stronglier than HC (suggesting that O_2 inhibits the reaction).

A very brief mechanism is proposed to discuss HC oxidation :

$$C_3H_8 + * \rightarrow (C_3H_6)* + H_2 \qquad (* = \text{metal in a reduced state})$$

$$(C_3H_6)* \rightarrow 3(CH_2)*$$

$$(CH_2)* + M\text{-}O \rightarrow M + (CH_2=O)*$$

$$M\text{-}O + (CH_2=O)* \rightarrow M\text{-}OH + (HC\equiv O^+)* \rightarrow HCOOH + M$$

$$HCOOH \rightarrow H_2 + \mathbf{CO_2}$$

In the presence of water, a partial oxidation is observed giving : $\mathbf{RCHO} + M + H_2O$

3.2.2 Alkene oxidation

Once again, propene oxidation is strongly dependent of the nature of the metal (Tab. III). Pd and Rh are more active than Pt [18].

Table III : Activity in propene oxidation as the function of the nature of the metal

Metal	Pd	Pt	Rh
TON (s^{-1})	0.56	0.35	0.95

In a stoechiometric mixture of propane and propene, propene is oxidized at lower temperature than propane which is in good agreement with the inhibiting effect of C_3H_6 adsorption on C_3H_8 activation (Fig. 8).

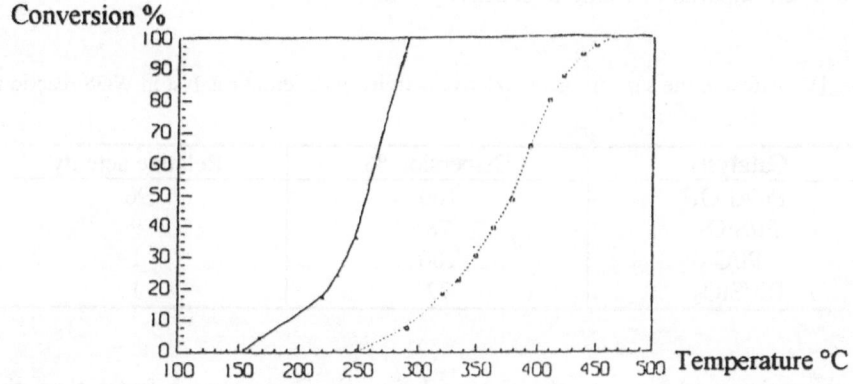

Figure 8 : Oxidation of a mixture C_3H_6/C_3H_8 on a Pt/Al_2O_3 catalyst (Dotted line : propane)

To conclude :

* Due to its strong interaction with the metal or its ability to oxidize it, oxygen may inhibit the reaction.
* Metal in its reduced state is more active than in its oxidized state. The simultaneous presence of oxidized metal and reduced metal is necessary to explain the redox mechanism.

4. ROLE OF WATER IN THE CO AND UNBURNT HYDROCARBONS COMBUSTION

In the exhaust gas, water is a product of important concentration (until 20% in volume). But steam is an oxidiser which can supply oxygen back during transient rich periods yielding CO_2 and H_2 by conversion of unburnt compounds (CO and hydrocarbons HC) [19].

The reactions involved are the Water Gas Shift and the steam reforming :

$$CO + H_2O \rightarrow CO_2 + H_2 \qquad\qquad \text{Water Gas Shift} \qquad\qquad (1)$$

$$C_3H_8 + 3H_2O \rightarrow 3CO + 7H_2 \quad (400\text{-}500°C) \quad \text{Hydrocarbon steamreforming} \quad (2)$$

$$3CO + 3H_2O \rightarrow 3CO_2 + 3H_2 \quad (200\text{-}300°C)$$

4.1 Water Gas Shift reaction

For the WGS reaction, Pt appears to be the best catalyst, the nature of the support is however an important parameter of activity (Tab. IV) [20].

Table IV : Effect of the support on the relative activity of different catalyst in WGS reaction

Catalysts	Dispersion %	Relative activity
Pt/Al_2O_3	100	90
Pt/SiO_2	78	9
Pt/C	100	1
Rh/SiO_2	72	1

Whatever the used catalyst, the CO oxidation by O_2 is always faster than the conversion with water.

So on Figure 9, the two reactions are carried out together in a O_2-H_2O mixture in an under stoichiometry of O_2 corresponding to 40% of the amount of O_2 required for a total conversion to CO_2. It appears that, until 40% of conversion, oxidation of CO by O_2 takes place and only after the WGS reaction occurs [19].

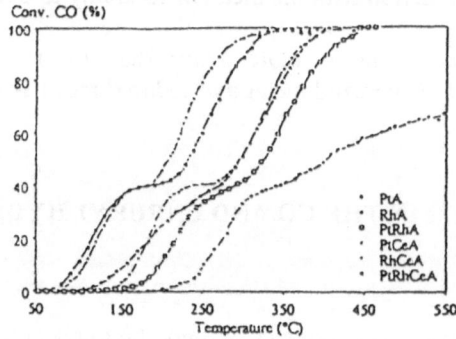

Figure 9 : CO conversion by a O_2-H_2O mixture in an under stoichiometry of O_2

Two kinds of WGS mechanism are proposed :

- Associating mechanism (on noble metal (like Pt)) :
CO + * → CO*
H_2O + 2* → OH* + H*   Dissociation of water
OH* + CO* → HCOO* + *
HCOO* + * → CO_2 + H*   Decomposition of the formiate species in CO_2
2H* → H_2 and H_2

- Regenerating mechanism (on easily oxidisable metal (Cu, Co...)) :

$H_2O + * \rightarrow H_2 + O*$ Surface oxidation by water

$CO + * \rightarrow CO*$

$CO* + O* \rightarrow CO_2$ Regeneration of the surface

4.2 Hydrocarbon steamreforming reaction

The hydrocarbon steamreforming reaction occurs in two steps (see equation (2)). The Pt-Rh bimetallic system appears to be the best catalyst : Rh being the more active for the first step (oxidation of HC to CO), and Pt for the second one (WGS reaction).

The steamreforming reaction remains interesting only in a rich mixture (under-stoichiometry of O_2), because the HC oxidation (propane oxidation for example) is always faster than steamreforming.

5. NO$_x$ REDUCTION

The catalytic conversion of nitrogen oxides has provided one of the most important scientific challenges of the last decades.

The most practical and convenient method for removing NO is catalytic reduction using unburnt compounds of the exhaust gas.

Two different practical conditions must be considered :

- **stoichiometric conditions**: the oxygen admission is controlled as to be just sufficient for a total oxidation of fuel (Three-Way Catalysts).

- **lean conditions**: the oxygen admission corresponds to an excess of 10 to 20% as regards to fuel supply (Diesel engine).

5.1. Conversion of NO$_x$ in Three-Way Catalysts (TWC)

Only noble metals can be used because of their ability to support oxidizing and reducing cycles (transient rich or lean periods). They are mainly supported on oxygen-storage supports, such as CeO_2 deposited on Al_2O_3.

5.1.1 Reduction of NO by CO

Reduction of NO by CO is one of the most important catalytic reaction occurring in catalytic converters to remove NO from the engine exhaust.

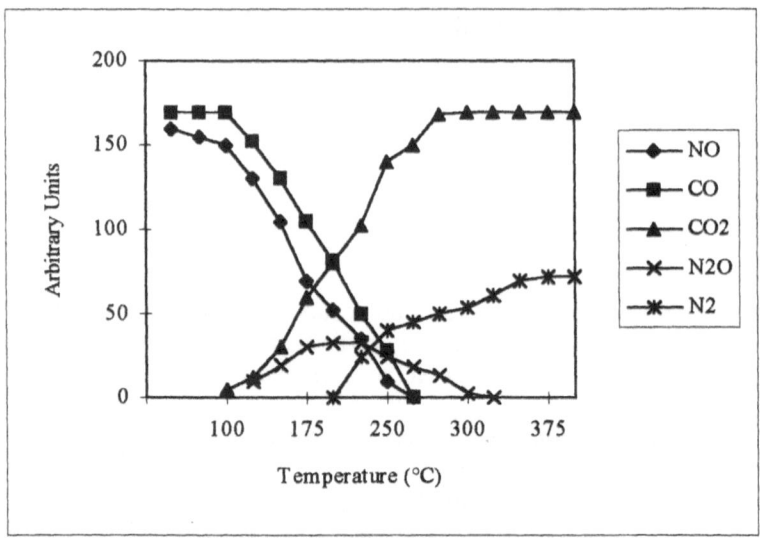

<u>Figure 10</u> : Typical curves of NO reduction by CO

Figure 10 shows a typical example of transformation curves, with the intermediate formation of N_2O which is a pollutant responsible for the green-house effect and for O_3 decomposition.

Rhodium is widely recognized as the most efficient catalytic component to promote the reaction of NO to N_2 in three-way catalysts.

In view of the historical developments in the understanding of the interaction of NO and CO with rhodium surfaces, HARRISON [21-22] has proposed a comprehensive mechanistic model of the NO + CO reaction as :

$$NO + * \Leftrightarrow NO* \qquad (1)$$
$$CO + * \Leftrightarrow CO* \qquad (2)$$
$$NO* + * \rightarrow N* + O* \qquad (3)$$
$$NO* + N* \rightarrow N_2 + O* + * \qquad (4)$$
$$NO* + N* \rightarrow N_2O* + * \qquad (5)$$
$$N_2O* \Leftrightarrow N_2O + * \qquad (6)$$
$$N_2O* \Leftrightarrow N_2 + O* \qquad (7)$$
$$2N* \rightarrow N_2 + 2* \qquad (8)$$
$$CO* + O* \rightarrow CO_2 + 2* \qquad (9)$$

Results indicate that the NO decomposition can make a major kinetic contribution in the NO + CO reaction system over supported rhodium catalysts.

At low temperatures, N_2O is mainly produced by reaction (5), while N_2 formation takes place at temperatures higher than 300°C.

The reaction rates for the NO reduction by CO are enhanced by an increase of the size of the rhodium particles, while the N_2 selectivity decreases.

5.1.2 Reduction of NO by H_2

H_2 can be present in the exhaust gas generated either by water-gas shift or steamreforming reactions.

Table V : Light-off temperatures for the reduction of NO by H_2 and CO

Catalysts	NO+H_2	NO+CO
Pt	121	471
Pd	106	431
Rh	163	296
Ru	237	205

Table VI : Selectivity of the reduction of NO by H_2

Catalysts	%Conv NO	Selectivity NO→N_2	NO→NH_3
Pt	94	23	77
Pd	94	26	74
Rh	100	67	33
Ru	100	92	8

Table V shows activity data obtained for the reduction of NO by H_2 over four supported noble metal catalysts [23]. The relative activity sequence is Pd>Pt>Rh>Ru. Similar data are shown for the reduction of NO by CO. In this case, the activity sequence is the reverse of that for the NO/H_2 reaction: Ru>Rh>Pd>Pt. This means that for Pt, Pd and Rh, the reaction NO with H_2 is faster than NO with CO.

On Pd and Pt, most of the NO is converted to NH_3 (Tab. VI). The Rh and Ru catalysts show much lower amounts of NH_3 formed. Actually by the volatility of the Ru oxides, Ru catalysts cannot be practically used. So Rh is the best metal for a good activity and selectivity.

5.1.3 Reduction of NO by unburnt hydrocarbons

The reduction of NO by alkanes under stoichiometric conditions has revealed that Pt is the most active species [23]. At the opposite, Rh is very usefull for the reduction of NO by olefinic compounds (Tab. VII).

It seems that this reaction is very sensitive to the structure of the catalyst and to the nature of the support. The incorporation of ceria to the alumina support, promotes the reduction of NO by HC. Its role as an oxygen storage component is manifested in the ability of ceria containing catalysts to store oxygen under lean operating conditions - thus promoting the conversion of NO_x - and release it under rich conditions by reaction with CO, H_2 or HC.

The mechanism proposed by Harrison [23] for the reduction of NO by CO is well accepted in the case of hydrocarbons as reductant.

Table VII : Reduction of NO by hydrocarbons - Effect of the nature of the catalyst

Metal	NO reduction by CH_4	NO reduction by C_3H_6
	Light-off temperature (°C)	Light-off temperature (°C)
Pt	330	380
Pd	420	320
Rh	450	300

5.2. Reduction of NO_x under lean operating conditions - Application to the diesel engine

The selective catalytic reduction of NO to N_2 by hydrocarbons in the presence of an oxygen excess is an important reaction in relationship with removal of NO_x from the exhaust gas of diesel and lean-burn engines.

Diesel and lean-burn gasoline engines generally operate under net-oxidizing conditions, typically at air/fuel ratios greater than 17. In such conditions, the level of unburnt compounds is very low and a further supply of fuel in the exhaust gas could be required.

5.2.1 General aspects

In an oxidizing atmosphere, noble metals or reductible oxides play an important role in the conversion of NO_x.
In these conditions, two reactions coexist :

$$NO + Red \rightarrow N_2 + Ox \qquad (1)$$

$$Red + O_2 \rightarrow Ox \qquad (2)$$

Actually the challenge of the catalytic research is to improve the first equation.

It was shown previously that catalytic oxidation of hydrocarbons (HC) is strongly dependent on :
  - **the nature of HC**
  - **the nature of the metal**
  - **the relation HC/Metal**

It appears that the most oxydisable hydrocarbons are the most effective for the selective NO reduction (Fig. 11) [24].

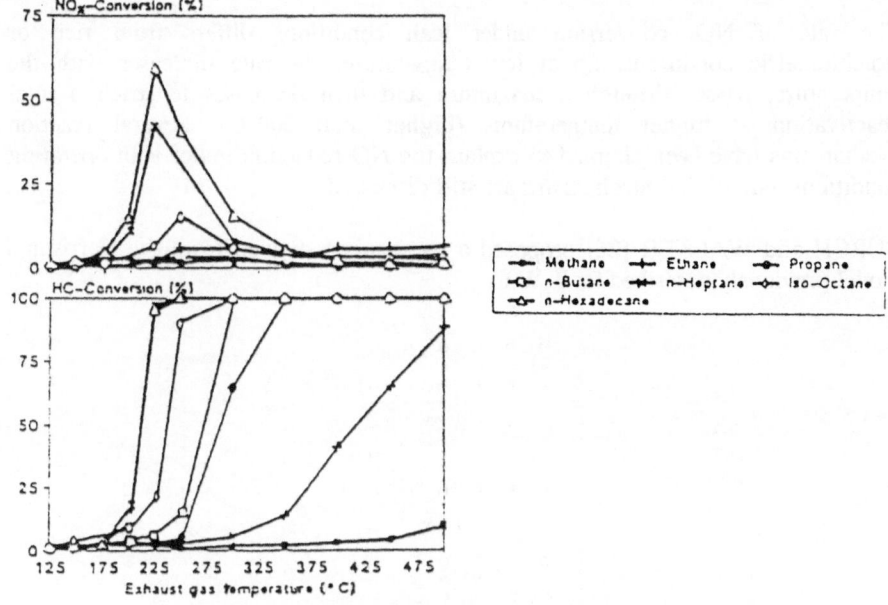

Figure 11 : HC and NO_x conversion as a function of the exhaust gas temperature for different hydrocarbons

Moreover BURCH [25] suggests that the NO reduction takes place as soon as the HC oxidation starts (Fig. 12).

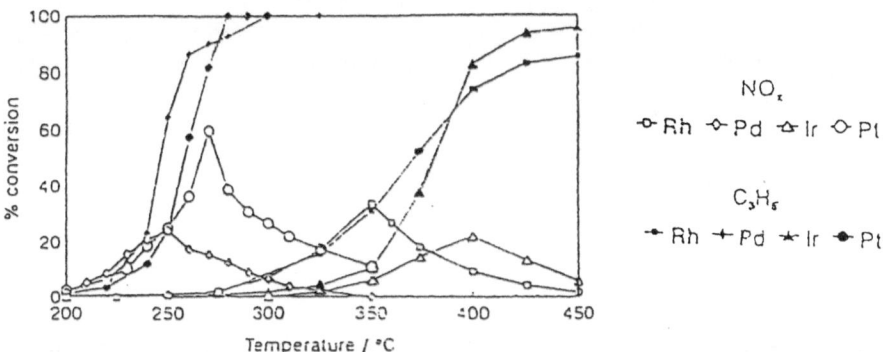

Figure 12 : Attendant NO_x reduction and C_3H_6 combustion on noble metal catalysts

To conclude, the most active catalysts for the HC oxidation, the best catalysts for the selective NO reduction.

5.2.2 Fundamental approach

The rate of NO_x conversion under lean conditions differs from rich or stoichiometric conditions. So at low temperature, the rate increases with the temperature, passes through a maximum and then decreases to reach a total deactivation at higher temperatures (higher than 500°C). Several reaction mechanisms have been claimed to explain the NO reduction under lean operating conditions, but all that mechanisms are still discussed.

BURCH and WALKER [26] proposed a mechanism derived from the Harrison's model previously described for T.W.C.

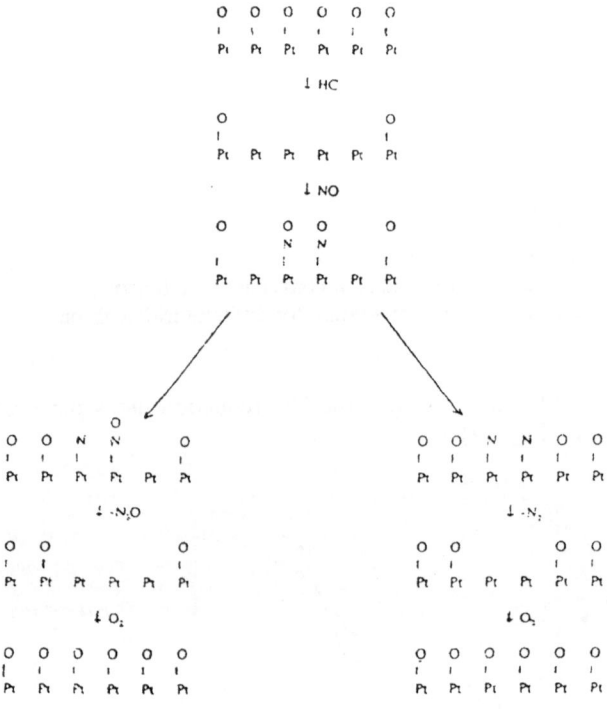

Scheme 3

In this mechanism, the hydrocarbon reduces a patch of Pt atoms from PtO to Pt metal. NO adsorption takes place on these Pt sites. At low temperature, dissociated and molecular adsorbed NO can coexist yielding N_2O formation. At higher temperature where the rate of NO dissociation is higher, leading to a higher concentration of N adatoms, it is expected that N_2, formed by N-N recombination, is the predominant product.

Concerning the maximum of conversion with temperature, TAYLOR suggested that the adsorption of NO cannot occur at high temperature. But the controversy is observed in TWC, for the same catalysts which are efficient at high temperature for NO reduction.

BURCH proposed that the difference with TWC could be a total oxidation of the catalyst surface in lean conditions. But it is well known that the oxide decomposition of noble metal can occur in the range of temperature considered, according to the equilibrium : $MO_x \Leftrightarrow M + x/2\, O_2$

On the other hand, WALKER argued that the loss of activity and so the maximum in conversion is in relationship with the complete disappearance of hydrocarbons.

Actually, on some solid catalysts like perovskite, the decreasing activity can be observed when a large amount of hydrocarbon remains unburnt. In fact, it is worthy of note that the catalytic activity decreases with the thermodynamic disappearance of NO_2 at high temperature. A lot of works proved the prevailing role of NO_2 in the catalytic reduction of NO_x in lean conditions.

So ENGLER et al. [24] proposed a mechanism involving surface reactions between NO_2 and oxygenated compounds as intermediate reactants. A hypothesis of a ''dual site'' reaction on supported precious metal catalysts was proposed (Fig. 13), and is rationalized by the following sequence of surface steps :

$$NO + [S_1] \longrightarrow NO \cdot [S_1]$$

$$NO \cdot [S_1] + 1/2\, O_2 \longrightarrow NO_2 \cdot [S_1]$$

$$NO_2 \cdot [S_1] \longrightarrow NO_2 + [S_1]$$

$$C_x H_y + z O_2 + [S_2] \longrightarrow C_x H_y O_{2z} \cdot [S_2]$$

$$NO_2 \cdot [S_1] + C_x H_y O_{2z} \cdot [S_2] \longrightarrow 1/2 N_2 O + x CO_2 + y H_2 O + [S_1] + [S_2]$$

$$NO \cdot [S_1] + C_x H_y O_{2z} \cdot [S_2] \longrightarrow 1/2 N_2 + x CO_2 + y H_2 O + [S_1] + [S_2]$$

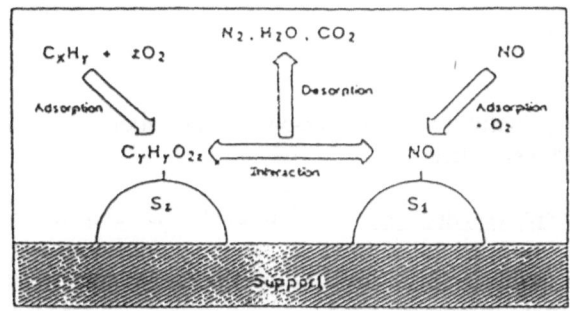

<u>Figure 13</u> : Proposed ''dual-site'' mechanism for NO_x reduction
in lean diesel exhaust gas on supported metallic catalysts

At present, the study of the surface organic chemistry, occuring in the NO_x conversion, is investing. Indeed, it is well accepted that the initial step of the reduction of NO is the oxidation of NO to NO_2. The subsequent steps leading to the formation of N_2 is the reaction of NO_2 with hydrocarbons to form organic nitro-compounds which are oxidatively decomposed to N_2 [27].

6. CONCLUDING REMARKS

The legislation on automotive pollution control will be stricter in the next years. The main improvements expected from heterogeneous catalysis are:

- better light-off from a cold start
- elimination of hydrogen sulfide release
- development of $DeNO_x$ catalytic converters for lean burn conditions.

For the three way catalysts, much progress have to be made for development of new catalytic phases allowing an improvement in activity (lower light off temperatures) and in lifetime (better stability at higher temperatures). On the other hand, the development of gasoline reformulation and the usage of new gasoline (with high H/C ratio in order to decrease the emission of carbon dioxide) will require new adsjusted catalysts.

Concerning the reduction of NO in lean conditions, the complexities of the various reactions involved prove that more data on mechanisms, reaction kinetics, equilibrium conditions, catalytic processes and overall chemistry are needed. It is obvious that in such experimental conditions the mechanisms involved are totally different than those proposed in stoichiometric conditions. The necessary solutions to $DeNO_x$ problems are related directly to additional understanding of the overall complex chemistry.

Acknowledgments

The authors are grateful to all the persons of the laboratory of whom the work allows us to write this course :

Permanent researchers : J. Barbier Jr., D. Duprez, P. Marécot, J. C. Ménézo, J. Rivière.
Students : T. Bertin, D. El Azami El Idrissi, A. Fakche, S. Inkari, B. Kellali, D. Martin, L. Pirault, R. Taha.

References

[1] Wei J., *Adv. Catal.*, **24** (1975) 57.
[2] Taylor K.C., Catalysis and Automotive Pollution Control, CaPoC 1 (Crucq A. and Frennet A., Eds) Stud. Surf. Sci. Catal., **Vol. 30**, Elsevier, Amsterdam, 1987. p.97.
[3] Taylor K.C., Catalysis, Science and Technology (Anderson J.R. and Boudart M., Eds), **Vol. 5**, Springer-Verlag, Berlin, 1984, p.119.
[4] Moro-Oka Y. and Ozaki A., *J. Catal.*, **5** (1966) 11.
[5] Moro-Oka Y., Morikawa Y. and Ozaki A., *J. Catal.*, **7** (1967) 23.
[6] Moro-Oka Y. and Ozaki A., *J. Catal.*, **12** (1968) 36.
[7] Ertl G. and Koch J., Z. Naturfarsh. Tul., **A25** (1970) 1906.
[8] Ertl G., Neumann M. and Streit K.M., *Surf. Sci.*, **64** (1977) 393.
[9] Weinberg W.H., Comrie C.M. and Lambert R.M., *J. Catal.*, **41** (1976) 489.
[10] Taylor J.L., Ibboston D.E. and Weinberg W.H., *Surf. Sci.*, **79** (1979) 349.
[11] Blyholder G.J., *J. Phys. Technol.*, **11** (1974) 845.
[12] Matshushima T., Almy D.B. and White J.M., *Surf. Sci.*, **67** (1977) 89.
[13] Engel T. and Ertl G., *J. Chem. Phys.*, **69** (1978) 1267.
[14] Bonzel H.P. and Ku R., *J. Vac. Sci. Technol.*, **9** (1972) 663.
[15] Yu Yao Y.F., *Ind. Eng. Chem. Prod. Res. Dev.*, **19** (1980) 293.
[16] Hicks R.F., Li H., Young M.L. and Lee R.G., *J. Catal.*, **122** (1990) 280.
[17] Garbowski E. and Primet M., *Appl. Catal.*, **125** (1995) 185.
[18] Yu Yao Y.F., *J. Catal.*, **87** (1984) 152.
[19] Barbier J.Jr. and Duprez D., *Appl. Catal.B : Environmental*, **4** (1994) 105.
[20] Grenoble D.G., Estadt M.M. and Ollis D.S., *J. Catal.*, **67** (1981) 90.
[21] Harrison B., Diwell A.F. and Hallet C., *Plat. Met. Rev.*, **32** (1988) 73.
[22] Harrison B., Wyatt M. and Gough K.G., Catalysis (Kemball C. and Dowden D.A., Eds), The Royal Society of Chemistry (London), **5** (1982) 127.
[23] Kobylinsky T.P. and Taylor B.W., *J. Catal.*, **33** (1974) 376.
[24] Engler B.H., Leyrer J., Lox E.S. and Ostgathe K., Catalysis and Automotive Pollution Control III, Studies in Surface Science and Catalysis, Elsevier, **96** (1995) 529.
[25] Burch R., Millington P.J. and Walker A.P., *Appl. Catal.B : Environmental*, **4** (1994) 65.
[26] Ansell G.P., Golunski S.E., Hayes J.W., Walker A.P., Burch R. and Millington P.J., Catalysis and Automotive Pollution Control III, Studies in Surface Science and Catalysis, Elsevier, **96** (1995) 577.
[27] Hamada H., *Catalysis Today*, **22** (1994) 21.

Chemisorption Bonds at Transition Metal Surfaces: Orbital Approach

D. Simon

*Laboratoire de Chimie Théorique, École Normale Supérieure de Lyon,
46 allée d'Italie, 69364 Lyon cedex 07, France
and
Institut de Recherches sur la Catalyse, CNRS, 2 avenue Albert Einstein,
69626 Villeurbanne cedex, France*

---

1. INTRODUCTION

This paper is devoted to the study of the orbital interactions occuring during processes involving a molecule and a metal surface. The typical case is the formation of a bond during chemisorption, but the study is also relevant to discuss other steps of catalytic reactions, like desorption, diffusion, dissociation of adsorbed molecules. We have intentionally omitted a presentation of the methods of calculations. The key concept, the local density of states, may be calculated from semi-empirical models [1], as well as from *ab initio* computations [2]. The analysis arising from these methods has the same basis, that is why we have avoided the technical aspects of the actual calculations. The examples used here, concerning mainly the chemisorption of a butadiene molecule on a Pd surface, have been calculated by an Extended Hückel method, which is simple enough to allow an easy introduction to concrete cases.

2. THE CONCEPTS

2.1 Definition of local density of states

The interaction between an adsorbed molecule and a metal surface may be understood in a way analogous to that of the interaction between molecular fragments. An overlap between the orbitals of the adsorbate and the metal surface occurs, leading to the formation of a bond, with a bonding or an antibonding character. So we need to characterize the molecular orbitals of the system, before and after the chimisorption.

The molecular orbitals of the piece of metal, with a naked or adsorbate covered surface, present two features, due to the large size of the system. First, they yield a continuous energetical spectrum, that we can represent as a density of states. Let us recall that the density of states, $n(E)$, counts the number of states between E and $E + dE$. Secundly, they expand on the whole piece of metal, appearing as a decomposition on the atomic orbitals of the whole sytem. How can we isolate, from this large set, the local properties associated with the interaction of the molecule and the surface site of the metal ? The answer lies in the calculation of the local density of states (ldos).

Let us take the case of an adsorbated molecule. We consider a particular molecular orbital, of energy E. We can say that, locally it presents a peculiar linear combination of the atomic orbitals of the adsorption site and the orbitals of the adsorbated molecule. This combination is the image of the interactions, coming from bonding or antibonding overlapping. Consequently, the orbitals are in phase or in opposite phase, respectively. What is, at this energy E, the relative contribution of each of these orbitals ? To answer this question, we have to make a partition, as that proposed by Mulliken [3], and widely used in orbital computations. This orbital $\Psi(E)$, at the energy E, is written as the following linear combination of φ_i, orbitals of the fragments (metal and adsorbated molecule) :

$$\Psi(E) = \sum_i c_i(E) \, \varphi_i \tag{1}$$

We give to each orbital φ_i, being either an atomic orbital, or a fragment orbital of the adsorbate, a partial occupation, at energy E :

$$p_i(E) = c_i(E)^2 + \sum_{j \neq i} c_i(E) \, c_j(E) \, S_{ij} \tag{2}$$

where S_{ij} is the overlap integral between the orbitals φ_i and φ_j.

How can we explain, physically, this spatial partition ? Let us take the example of an orbital φ_2 of the adsorbated molecule, interacting with an atomic orbital, φ_1 of the surface (Scheme 1). The overlap area, indicated by an arrow correponds to a weight $2 \, c_1 c_2 S_{12}$. Each orbital has the $c_1 c_2 S_{12}$ term, accounted for its occupation. The separation is represented by a dashed line in Scheme 1. This means, that, the overlap term has been separated in two equal parts, one part

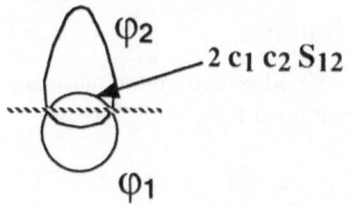

Scheme 1

for each orbital. Of course, this partition is somewhat arbitrary, but it has the advantage to preserve the normalization, i.e.

$$\sum_i p_i(E) = 1 \qquad (3)$$

and it does not give any weight to one of the interacting orbitals. It can be also applied to the naked surface orbitals, for which $p_i(E)$ contains the overlap terms coming from the other atomic orbitals of the surface.

So, we can assign to each orbital, φ_i, a local density of states (ldos) [4], defined as the density of states of the system, weighted by the $p_i(E)$ population, at the energy E[1]. This concept of ldos is quite fundamental in the understanding of the orbital interactions between a molecule and a metal surface. It allows the decomposition of the density of states in local components, leading to a formalism analogous to that of pure molecular interactions. Nevertheless, the surface problem remains fundamentally different, in that sense that the energetical spectrum is continuous. The ldos depicts the energetical spectrum of the states associated, locally, with an orbital. Now, let us consider the quantities that can be evaluated from the ldos.

---

[1] A more formal definition of the density of states is the following [5] :

$$n(E) = \sum_\alpha \delta(E-E_\alpha)$$

where E_α are the eigenvalues of the system, and δ the Dirac function.
The local density of states is then defined by :

$$ldos(E) = \sum_\alpha p_i(E_\alpha) \, \delta(E-E_\alpha)$$

2.2 Electron population and transfer

It is easy to calculate, from the ldos associated with an orbital φ_i, the total occupation[2], N_i, up to the Fermi level E_f :

$$N_i = \int_{E_{min}}^{E_f} ldos(E)\, dE \tag{4}$$

The electron tranfers are evaluated by the balance between the loss and the increase in electronic population of each orbital.

2.3 Mean energy

As the contribution of an orbital φ_i at energy E is known, it is possible to calculate the total energy of the electrons filling this orbital :

$$E_T = \int_{E_{min}}^{E_f} E\, ldos(E)\, dE \tag{5}$$

Then, it allows the calculation of the mean electron energy associated with φ_i :

$$< E_i > = \frac{E_T}{N_i} \tag{6}$$

This energetical value corresponds to the mean energetical position of the orbital φ_i, i.e. to what is often called the "centre" of the occupied band, associated with φ_i. $< E_i >$ is a measure of the stability of φ_i, interacting with the system. Obviously, during the bond formation, one can estimate the evolution of the energy of the orbital φ_i, undergoing either a stabilization when $< E_i >$ decreases, or a destabilization when $< E_i >$ increases. Let us emphasize that, as the ldos function, $< E_i >$ is based on a peculiar partition of space, generally of Mulliken type.

2.4 Overlap population

In the case of molecular systems, the bond may be characterised by the overlap population between the interacting orbitals. The same concept is valid for the metal surface systems : at the energy E, the $2\, c_i(E)\, c_j(E)\, S_{ij}$ term is a signature of the bond between two orbitals φ_i and φ_j. When this term is positive, the

---

[2]The local density of states as defined above is normalized to 1, so the total occupation of the orbital , with 2 electrons, would correspond to a value of 1 for the integrated ldos

interaction between φ_i and φ_j is bonding, and when the term is negative, it is anti-bonding. A non-bonding contribution is associated with a low value of this overlap term. By weighting the energy level E by $2 c_i(E) c_j(E) S_{ij}$, an overlap population density of states, opdos, may be defined[3] [6]. The sum of the opdos up to the Fermi level gives the total overlap density associated with the two orbitals φ_i and φ_j.

2.5 Atomic properties

These properties, ldos, N_i, $< E_i >$ are associated with an atomic or a fragment orbital. They can be cumulated for all the orbitals of an atom, or a molecular fragment, giving then values relative to the whole species considered. In the same way, the opdos between two atoms is evaluated by cumulating the opdos associated with all pairs of φ_i and φ_j orbitals belonging to the two atoms.

3. ILLUSTRATION OF THE CONCEPTS

Scheme 2

Let us examine the case of the adsoprtion of a butadiene molecule on the (111) surface of a Pd cluster (Scheme 2). The Pd cluster has a sufficient size, 114 atoms in the example of Scheme 2, so that its surface simulates properly an infinite (111) surface of a piece of Pd metal [7]. The adsorption is a di-Π one [8]. It means that the interaction between the butadiene molecule and the surface occurs with two adjacent Pd atoms, and mainly via the Π system of the butadiene molecule. These Π orbitals are constituted of bonding, occupied orbitals, 1Π and 2 Π, and antibonding, unoccupied orbitals 1Π^* and 2Π^* (Figure 1). They lie at energies of the same scale of the metallic band, as illustrated by Figure 1. Let us describe the nature of the Pd atomic orbitals that interact with the butadiene molecule. For a sake of simplicity, we look only at the d orbitals. As illustrated in

---

[3] In the case of periodic systems, R.Hoffmann has defined the Crystal Orbital Overlap Population (COOP) which is equivalent to the opdos.

D. Simon

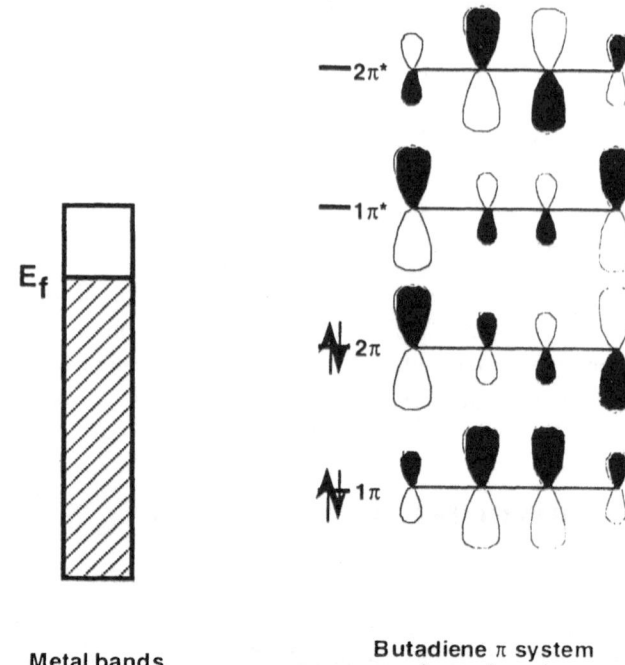

Metal bands

Butadiene π system (gas phase)

Figure 1. Occupied and unoccupied orbitals of the butadiene molecule. The energy scale of the metal band is given for comparison.

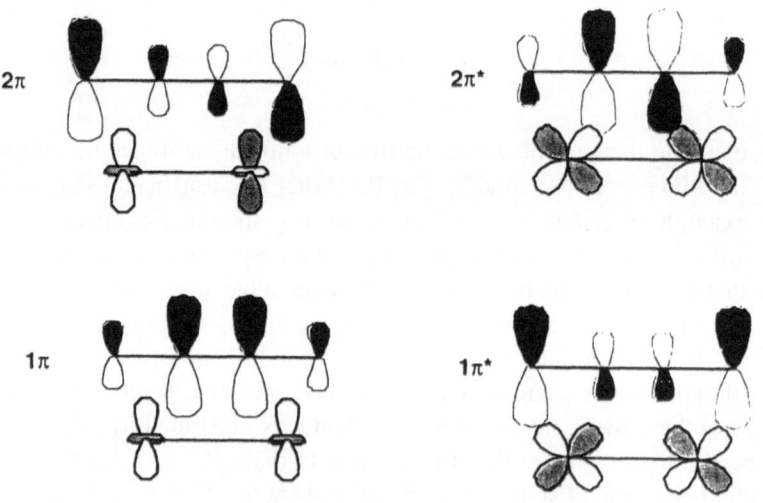

Figure 2. Interaction of 1Π and 2Π with combinations of d_{z^2} orbitals of the Pd_2 site; interaction of $1\Pi^*$ and $2\Pi^*$ with combinations of d_{yz} orbitals of the Pd_2 site

Figure 2, for symmetry reasons, 1Π and 2Π interact, with the d_{z^2} orbitals of the Pd atoms, whereas $1\Pi^*$ and $2\Pi^*$ interact with d_{yz}, and d_{xz} orbitals.

What is the shape of the ldos of the surface Pd atoms under interaction ? More precisely, Figure 3 depicts the ldos associated with d_{z^2}, and d_{yz} orbitals, before and after the butadiene adsorption. We see clearly the modification induced in the ldos : the general shape of the "d-band" between -12 and -10 eV is strongly perturbated, and peaks appear, at low energy (around -12.5 eV) in the d_{z^2} case, and around -7.5 and -8.5 eV in the d_{yz} curve. These additional peaks come from the interaction with the Π and the Π^* orbitals, respectively. Indeed, we can find again these peaks in the Π and Π^* ldos (Figure 4). Moreover, the Π and Π^* ldos show the interaction of these butadiene orbital with the metal band, since contributions appear in both cases in the -12 to -10 eV region.

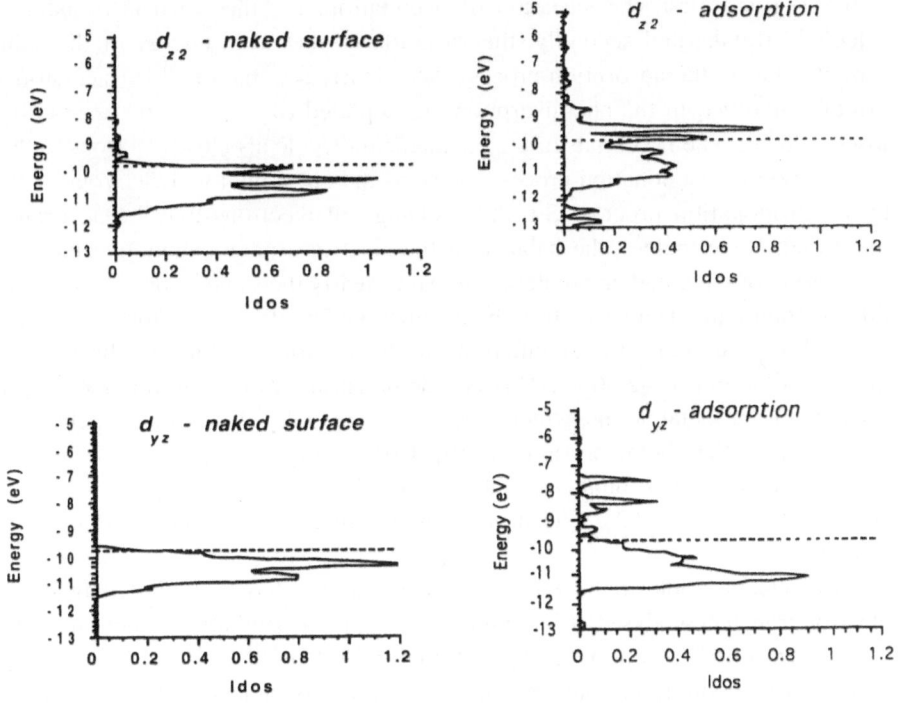

Figure 3. ldos of d_{z^2} and d_{yz} before adsorption (naked surface) and after butadiene adsorption. The dashed line corresponds to the Fermi level.

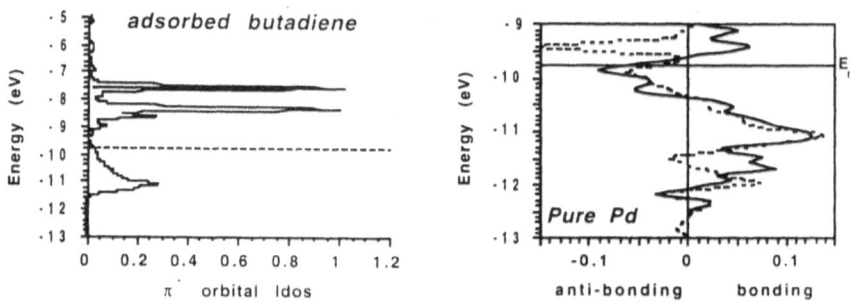

Figure 4. Ldos of Π and Π* orbitals under butadiene adsorption. The ldos are cumulated over 1Π and 2Π (Π curve), and 1Π* and 2Π* (Π* curve). The dashed line indicates the Fermi level.

In order to illustrate the use of total population, and of mean energy calculations, extracted from the integration of the ldos up to the Fermi level, Table 1 gives, first, the variation of occupation, i.e. the electron transfer of selected orbitals, and secondly, the variation of the mean energy of the same orbitals, during the adsorption process. We clearly see that the d surface atomic orbitals involved in the chemisorption are depleted (d_{z^2} loses $-.62$ e⁻, and d_{yz} loses $-.38$ e⁻). The Π orbitals of the butadiene molecule also lose electrons ($-.90$), it corresponds to a donation process, whereas the Π* orbital win electrons ($+.96$) by a retrodonation process. So, the exchange of electrons may be accurately identified. Nevertheless, the balance of the electron tranfers show that the total occupation of the butadiene molecule remains nearly the same ($+.06$ e⁻), when the Pd site population decreases strongly, as illustrated by the occupation variation of d_{z^2} and d_{yz}. Actually, the cumulation on all the atomic orbital of the Pd_3 site gives a depletion of -2.30 e⁻. This is a demonstration of the importance of the remains of the metal atoms, which play a reservoir role, able to receive electrons coming from the chemisorption site [9]. Concerning the mean energy, we see (Table 1) that all the orbitals are stabilized by the chemisorption process. This is consistent with the -1.02 eV adsorption energy, that we have calculated using the same model [7].

Now, let us look at the Pd_2 site, since, as we have concluded from the electron transfers analyses, it is submitted to a large perturbation under butadiene chemisorption. The opdos curves of the Pd–Pd bond (Figure 5) show the evolution of bonding and anti-bonding during the butadiene chemisorption. At first sight, the perturbation seems moderate. Nevertheless, we see that the bonding contribution around -11.5 eV disappears during chemisorption. This phenomenon comes from the interaction with the occupied orbitals of butadiene : 1Π and 2Π

Table I. Electron transfer Δp and variation of mean electron energy, $\Delta<E>$, during butadiene chemisorption. The d_{z^2} and d_{yz} orbitals are those of the Pd site; Π and Π^* represent the orbitals of the butadiene molecule. Δp is cumulated, for Π and Π^*, on the two corresponding orbitals 1Π, 2Π and $1\Pi^*$, $2\Pi^*$, respectively. $\Delta<E>$ is detailed for 1Π and 2Π, but cannot be evaluated for $1\Pi^*$ and $2\Pi^*$ since they are empty for butadiene in the gas phase.

Orbital	Δp (e^-)	$\Delta<E>$ (eV)
d_{z^2}	−0.62	−0.34
d_{yz}	−0.38	−0.31
Π	−0.90	1Π : −0.19 ; 2Π : −0.44
Π^*	+0.96	

are situated in this energy zone in the gas phase. It is related to a lowering of the bond strength of the Pd_2 adsorption site. A perturbative effect is also observed at −9.5 eV since an antibonding contribution emerges, replacing a bonding peak. This contribution has no consequence for the Pd–Pd bond, since it is unoccupied, nevertheless, it is a mark of the interaction with the Π^* butadiene orbitals.

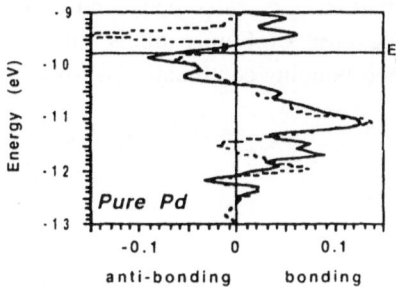

Figure 5. Opdos of the Pd–Pd bond of the adsorption site : full line : before adsorption ; dashed line : after adsorption.

4. INTERACTION DIAGRAMS

The analysis of the ldos curves may be understood in a global manner by the use of diagrams, illustrating the main processes occuring during the orbital interaction. We can distinguish the interaction between occupied and vacant orbitals, or between two occupied orbitals. Moreover, in this particular case of metallic system, we shall see that the interaction between two vacant orbitals,

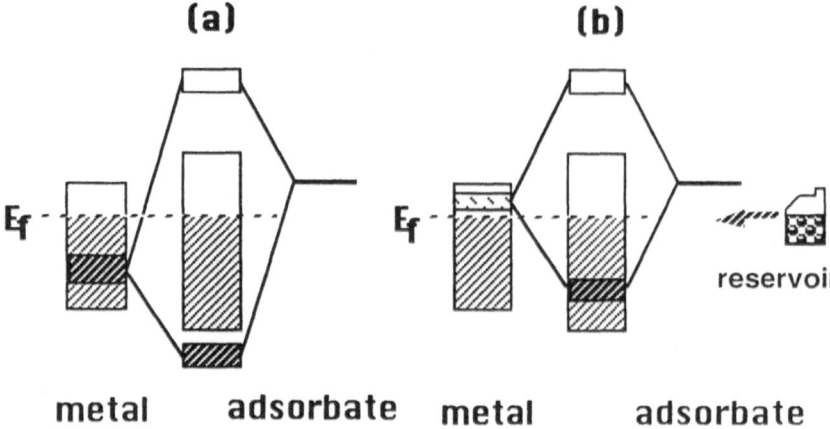

Figure 6. Interaction diagram between an adsorbate and the metal surface : (a) interaction between an occupied contribution of the metal and a vacant orbital of the adsorbate ; (b) interaction between two unoccupied contributions.

although without true physical sense, since no electron are present, may induce occupied contributions [5].

Figure 6a shows the stabilization coming from a 2-electron interaction, involving an occupied contribution and a vacant orbital. It is clear that the low energy combination has a bonding character and corresponds to a lowering of the global energy of the system. Of course, as in a molecular case, the electrons initially located on one partner, are shared between the two fragments, leading to an electron tranfer. The case of Figure 6b is more puzzling : two vacant contributions interact, their bonding combination is lowered, and passes below the

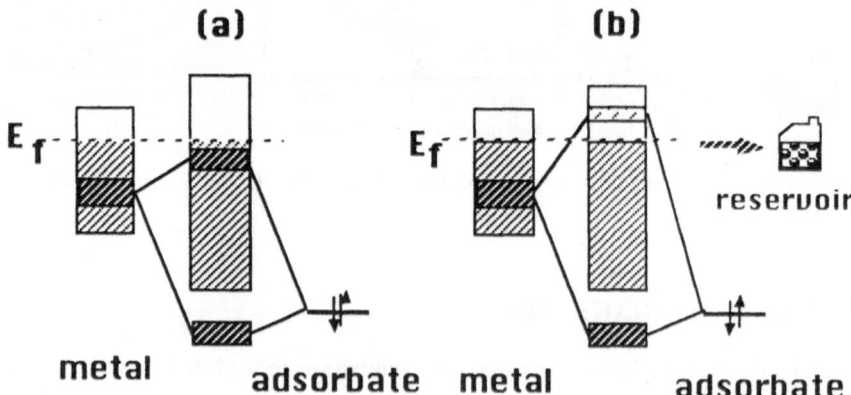

Figure 7. Interaction diagram between two occupied contributions. The resulting antibonding combination is : (a) below the Fermi level ; (b) above the Fermi level.

Fermi level. Then, it becomes occupied, by a transfer of electrons from the reservoir. As the lowering is going on, the stabilization becomes more end more effective.

Figure 7 depicts the interaction between two occupied contributions. First, in Figure 7a, the interaction leads to a bonding and an antibonding combinations, and the energetical result is destabilizing, since the antibonding character is stronger than the bonding one. As the interaction increases, the antibonding contribution may pass through the Fermi level [10] : it becomes unoccupied, the electrons being transferred to the reservoir (Figure 7b). The global energetical behaviour is then stabilizing.

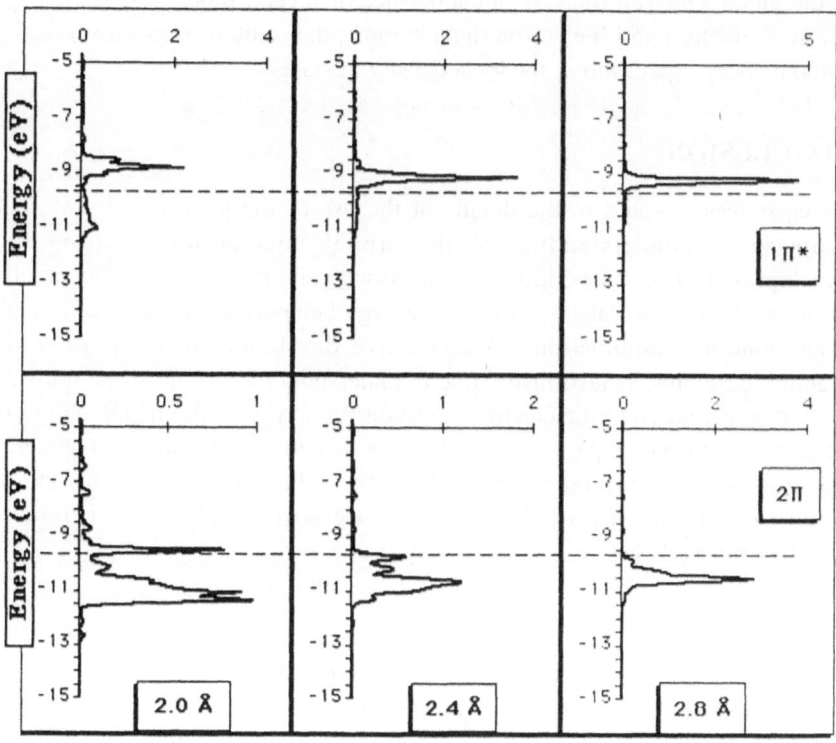

Figure 8. Illustration of the interaction diagram in the butadiene adsorption : the ldos of 2Π and 1Π* are represented as functions of the distance butadiene–surface.

Let us illustrate this behaviour in the case of butadiene adsorption on the Pd₂ site described above. In Figure 8, we show how the ldos, associated with 2Π and 1Π*, are modified as the butadiene surface distance decreases. At 2.8 Å, 2Π and 1Π* are nearly unperturbated, and the main peak corresponds to their energy in the gas pase. At 2.4 Å, 2Π has got contributions in the whole metal band, and a peak just below the Fermi level appears, corresponding to an antibonding combination. The $1\Pi^*$ is less perturbated, but contributions at low energy begin to emerge. At 2.0 Å, the antibonding contribution of the 2Π curve, is now above the Fermi level. The main peak of the $1\Pi^*$ orbital is located at higher energy, indicating an antibonding interaction, whereas the bonding combinations appear below the Fermi level.

Two points must be emphasized, that give the features of the molecule-metal surface interaction :

- the reservoir role of the metal : it allows the transfers of electron, being able to give or receive electrons during the chemisorption ; likewise, the Fermi level is held fixed by the reservoir.

- the global electron transfers are a balance of several processes, involving 2-electron, 4-electron and 0-electron interactions ; these interactions are spread on the whole energy spectrum of the local density of states.

5. CONCLUSION

This paper was devoted to the details of the use of the local density of states concept in the understanding of the orbital interactions occuring in a chemisorption procress. The ldos, associated with an orbital, allows quantitative evaluation of electron transfers, of mean energy behaviour, and of local bonding and antibonding combinations. A qualitative discussion, in terms of orbital interaction diagrams, is also instructive to understand the origin of the transfers. This matter is related to the electron reservoir properties of the metal, and to the frontier contributions, near the Fermi level. It would be relevant to consider, in connection with this orbital point of view, the global energetic behaviour split into attractive (2-electron) [11], repulsive (4-electron) [11], and internuclear repulsion terms [7,12].

References

[1] Baetzold R.C. in *Theoretical Aspects of Heterogeneous Catalysis*, Moffat J.B. Ed. (Van Nostrand Reinhold, 1990) pp. 458-505
Blyholder G. in *Quantum Chemistry Approaches to Chemisorption and Heterogeneous Catalysis* , Ruette F. Ed. (Kluwer, 1992) pp. 181-200
[2] Hammer B. and Nørskov J.K., *Nature* **376** (1995) 238-240
Van Santen R.A. and Neurock M., *Catal. Rev. Sci. Eng.* **37** (1995) 557
Paul J.F. and Sautet P., *Phys. Rev. B* **53** (1996) 8015-8027
[3] Mulliken R. S., *J. Chem. Phys.* **23** (1955) 1833
[4] Hoffmann R., *Solids and Surfaces : A Chemists's View of Bonding in Extended Structures* (VCH, 1988)

[5] Desjonquères M.C., Jardin J.P. and Spanjaard D., *J.Chim.Phys.* **85** (1988) 827-845

[6] Hoffmann R., *Rev. Mod. Phys.* **60** (1988) 601-628

[7] Hermann P., Simon D. and Bigot B., *Surf. Science.* **350** (1996) 301

[8] Hermann P., Simon D., Sautet P. and Bigot B., *J. Catal.* (1997) to be published

[9] Simon D. and Bigot B. *Surf. Science* **306** (1994) 459-470

[10] Simon D. *Thèse de Doctorat* (1990, Orsay)

[11] Delbecq F. and Sautet P. *Surf. Science* **295** (1993) 353-373

Delbecq F. and Sautet P., *J. Catal.* **152** (1995) 217-236

[12] Anderson A. B. J., *J. Chem. Phys.* **60** (1974) 2477

Savary F., Weber J. and Calzaferri G., *J. Chem. Phys.* **97** (1993) 3722

Characterization of Metallic Catalysts by X-Ray and Electron Microscopy Techniques

G. Bergeret

Institut de Recherches sur la Catalyse, CNRS, 2 avenue Albert Einstein, 69626 Villeurbanne cedex, France

This lecture is intended to describe briefly the X-ray and electron microscopy techniques and to give a survey of their application to the study of metallic catalysts.

The first section deals with relations between particle size distribution, mean diameters, dispersion, coordination number.

The second section - X-ray techniques - are divided in three parts, namely: diffraction (XRD), small-angle scattering (SAXS), and absorption spectroscopy (EXAFS, XANES). *In situ* phase identification, determination of the composition of supported bimetallic catalysts, line broadening analysis (LBA) are discussed in the XRD section. Although not very frequently used, the advantages of the SAXS will be emphasized. The application of EXAFS to the characterization of supported monometallic catalysts (study of the precursor preparation) is given. The characterization of bimetallic catalysts by EXAFS as well as by XANES is treated.

The third section, devoted to electron microscopy, presents for a non-specialist i) the different types of microscopes (the scanning probe microscopies will not be discussed here), ii) the preparation of samples, iii) the different types of images. Analytical microscopy - EDX in current use to determine the composition of bimetallic catalysts, EELS for structural information - is also presented.

1. DEFINITIONS and GENERALITIES

1. 1 Particle Size: Mean Diameters

Consider a collection of n_i spherical particles of diameter d_i. Several curves can be plotted against the diameter. The plot of n_i as a function of d_i gives the number distribution and leads to the length-number mean diameter $d_{LN} = \Sigma n_i d_i / \Sigma n_i$. The area distribution $n_i d_i^2$ versus d_i gives the volume-area mean diameter

$d_{VA} = \Sigma n_i d_i^2 d_i / \Sigma n_i d_i^2$. The volume-weighted mean diameter $d_V = \Sigma n_i d_i^3 d_i / \Sigma n_i d_i^3$ corresponds to the volume-weighted distribution $\Sigma n_i d_i^3$ vs. d_i. The area distribution and the volume-weighted distribution to a greater extent, give more weight to the larger particles. For any distribution of particle sizes, d_V will be larger than d_{VA} and d_{VA} will be larger than d_{LN}. The volume-area mean diameter d_{VA} is the most useful data for catalysis because it is related to the specific surface area. It is the parameter which is derived from gas adsorption measurements. Other techniques such as X-ray diffraction line broadening analysis (LBA), small-angle X-ray scattering (SAXS) and of course granulometry measurements by electron microscopy allow to obtain mean particle size. However, these techniques do not give the same type of average. Therefore, it should be specified which particle size average is measured depending on the experimental technique employed.

1. 2 Model Particles: Total Number of Atoms and Size

The total number of atoms as a function of particle size for spherical platinum particles is shown in Fig. 1. The curve indicates that the number of atoms increases very rapidly with particle diameter. More realistic particle models are obtained by considering polyhedra rather than spheres. Figure 1 gives the number of atoms for two fcc models: truncated octahedra [1] (often improperly called cubo-octahedra), and icosahedron [2, 3]. Theoretical predictions [3] and experimental data [4] for fcc metals show that the particles have fcc truncated octahedron shapes. Smaller particles containing less than 150-1000 atoms could be icosahedral. There is a good agreement between the spherical macroscopic model calculated from the atomic mass and the mass density and the crystallographic models (octahedra, cubo-octahedra, icosaedra...) where the positions of the atoms are well defined.

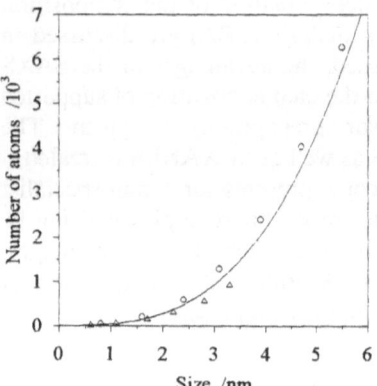

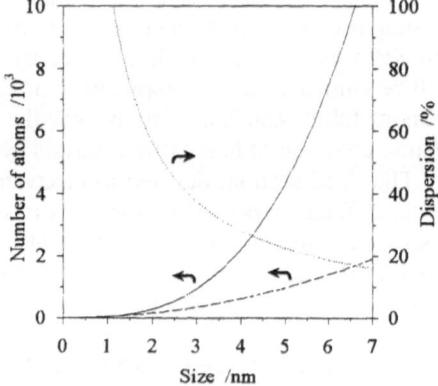

Figure 1: Number of atoms in model platinum particles as a function of particle size d. Continuous curve = total number of atoms in a sphere (ratio $(\pi d^3/6)/v$ where v is the volume of a Pt atom in bulk platinum; o = truncated octahedron; Δ = icosahedron.

Figure 2: Total number of atoms (continuous line), number of surface atoms (dashed line), and dispersion (dotted line) as a function of particle diameter d_{VA} for platinum (spherical model).

1. 3 Dispersion. Surface Area

The metal dispersion D is given by

$$D = N_S / N_T \tag{1}$$

where N_S is the number of metal atoms present on the surface of the particles and N_T the total number of metal atoms (surface and bulk). The dispersion. i.e. the fraction of surface atoms. is between 0 and 1 (or 0 and 100 %).

Relationships between dispersion, surface area and mean particle diameter can be established by making assumptions on the nature of the crystal planes exposed on the particle surface [5]. Thus, for a fcc metal, it is commonly assumed equal proportions of the (111), (100) and (110) planes. Therefore, it is easy to calculate the number of atoms per unit area (1.24 atom per 10^{-19} m^2 for Pt) and then the surface area a occupied by an atom on the particle surface (8.07 $Å^2$ for Pt).

The relationships between specific surface area (S_{SP}), dispersion (D) and mean particle size (volume-area mean diameter d_{VA}) are

$$S_{SP} = a (N_A / M) D = 6 / (\rho d_{VA}) \tag{2, 3}$$

where N_A is the Avogadro number. M the atomic mass and ρ the mass density.

The relationship between metal dispersion (D) and mean particle size is

$$D = 6 (v/a) / d_{VA} \tag{4}$$

where v is the volume occupied by an atom in the bulk of the metal ($v = M / \rho N_A = 15.1$ $Å^3$ where M is the atomic mass, ρ the mass density, and N_A the Avogadro number). This is illustrated in Fig. 2 for platinum. For fcc metals, with particle size d expressed in nanometers. Eq. 4 becomes

$$D = 1.1 / d \tag{5}$$

within 10 %. It should noted that the relationships 2-5 apply to spherical particles (or particles which can be assimilated to spheres) and the values obtained for the dispersion are only approximate because of the assumptions made on the number of atoms on the particle surface. For very small particles (< 1.2-1.5 nm). the macroscopic spherical model should not be used and crystallographic models assuming a given particle morphology should be considered.

1. 4 Coordination Number

The coordination number (CN) of a atom is the number of nearest neighbours, i.e. the number of atoms in the first coordination shell. The coordination number is higher for an atom in the bulk of the particle than an atom on the surface. For a fcc metal. CN is 12 in the bulk and 9. 8 or 7 for a surface atom on the low-index planes (111), (100), and (110) planes, respectively. For a atom on the edges or on the corners of the particle, CN is even lower. Therefore, the mean CN of a large fcc particle tends to 12 and decreases as the particle becomes smaller. The value of the mean CN gives information on the size of the particles (if a given morphology is assumed). Figure 3 gives the mean CN calculated for icosahedra and truncated octahedra as a function particle diameter for platinum.

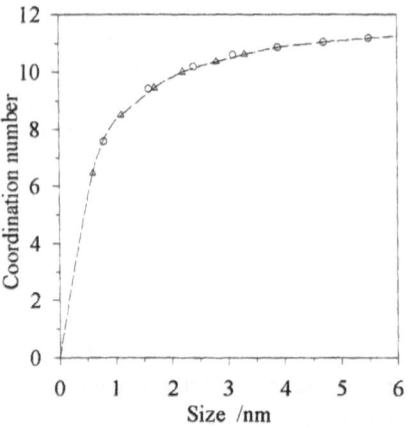

Figure 3: Plot of the mean coordination number as a function of particle diameter for platinum: icosahedra (Λ) and truncated octahedra (o).

2. X-RAY TECHNIQUES

There are many books on X-ray techniques. The fundamental aspects and the basis of powder diffraction are given in the still essential book by Klug and Alexander [6]. A recent description of the methods currently in use for powder diffraction can be found in a cheap but complete *"Course Notes"* [7]. Concerning catalytic materials, a review article by Gallezot [8] gives a complete survey of the application of X-ray methods to the study of catalysts. In the *"Handbook of Heterogeneous Catalysis"*, a section is devoted to powder diffraction [9] and to particle size measurements by different techniques including X-ray and electron microscopy techniques [5].

2. 1 X-Ray Powder Diffraction (XRD)

2. 1. 1 *Phase Identification*

X-ray powder diffraction is a standard tool for characterizing metallic catalysts. Phase identification is based on the comparison of the set of reflections with that of pure reference phases, or usually with a database. The Powder Diffraction File (PDF) [10], which contains 64 000 patterns with 12 000 concerning metals and alloys in 1997, is commonly used. Search programs (for example [11]) running on personal computers give a list of candidate phases in a few seconds. The conventional practice of phase identification is to study the catalyst in air at successive stages of its life (preparation, activation, after reaction). But *in situ* characterization under controlled atmosphere and temperature is a necessity insofar as the phases present in the catalyst can be air sensitive or deeply modified by the temperature and the reactive gases. For instance, Hardacre *et al.* [12] studied the preparation of Pt/ceria catalysts derived from Pt-Ce crystalline alloys. Figure 4 shows *in situ* XRD patterns of Pt_3Ce_7 precursor after treatment in 50 bar H_2 at different temperatures.

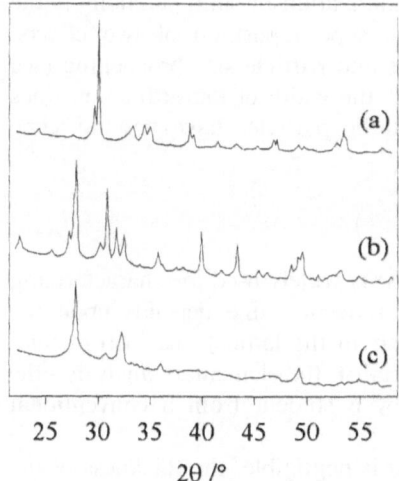 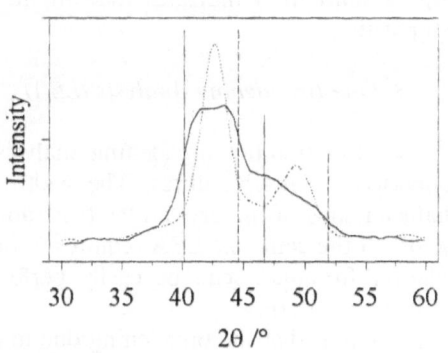

Figure 4: *In situ* XRD traces following the activation of Pt_3Ce_7 in 50 bar hydrogen at (a) 300 K, showing the alloy phase, (b) 520 K, showing the intermetallic hydride, and (c) 720 K, showing the formation of CeH_{2+x} [12].

Figure 5: *In situ* X-ray pattern of a Pd-Ni bimetallic catalyst after reduction at 490 K (solid line) and 720 K (dotted line) (40/60 nominal atomic composition) [14]. Vertical bars indicate the position of the (111) and (200) Bragg peaks for Pd (solid line) and Ni (dashed lines) extracted from the PDF [10].

Reaction with hydrogen at 720 K led to formation of a nonstoichiometric hydride CeH_{2+x}. An intermediate phase, thought to be an intermetallic Pt-Ce hydride was obtained at 520 K. Note that, after activation at 720 K, platinum was invisible to X-ray diffraction, probably because it was in a state of very fine division. Disappearing of the CeH_{2+x} hydride phase was rapid with air reoxidation at 370 K, forming CeO_2 and XRD-visible Pt particles. Oxidation of the intermetallic hydride leading to ceria and Pt particles, occurred at room temperature. It must be emphasized that these hydrides were observed because the samples were protected against the air.

Dynamic studies on a time scale of seconds are possible owing to the synchrotron radiation which allows to obtain a XRD pattern within a few tens of seconds [13].

2. 1. 2 Solid Solution Identification

When two alloy metals form a solid solution, the reflections of the solid solution are located between those of the two pure metals. Generally, the change of the unit cell parameter is linear or almost linear with the atomic concentration (Végard's law) and it is possible to determine the composition of the solid solution. Figure 5 shows the formation of Pd-Ni bimetallic particles supported on SiO_2 [14]. After reduction at 720 K, the (111) and (200) diffraction lines gives a Pd concentration in the bimetallic particles close to the value obtained by chemical analysis (36 and 40 %, respectively). But a problem with the characterization of solid solutions is the broadening of the diffraction lines due to the superimposition of two effects: distribution of the composition of alloy particles and particle size broadening (see the next section). After reduction at only 490 K, the width of the diffraction lines (Fig. 5 solid line) indicates that all the bimetallic particles have not the same composition.

2. 1. 3 Line Broadening Analysis (LBA)

X-Ray diffraction line broadening analysis (LBA) is widely used for characterizing supported metal crystallites. The width of an reflection line depends upon the characteristics of the crystallites (size and defects in the lattice) and instrumental factors. If the complete LBA requires X-ray specialist, the elementary analysis - the Scherrer formula - can be easily performed by a student from a conventional diffraction pattern.

Assuming that the broadening due to defects is negligible, the thickness of the crystallite in the direction perpendicular to the planes (*hkl*), L_{hkl}, is inversely proportional to the breath of the diffraction line and is given by the Scherrer formula

$$L_{hkl} = \lambda \, / \, (\beta_i \cos \theta_{hkl}) \tag{6}$$

where λ is the X-radiation wavelength employed, β_i the integral breadth (total area under the line profile divided by the line intensity at maximum) expressed in radians and θ_{hkl} the angular position of the peak maximum. Note that L is a volume weighted average dimension $L = \Sigma \, n_j L_j^3 \, L_j \, / \, \Sigma \, n_j L_j^3$. Before to calculate the area of the line, a good practice is to strip the $K\alpha_2$ line and to remove the pattern background using computer programs integrated with the diffractometers. Then, the calculated breadth must be decreased by the broadening due to the instrument estimated from the line breadth of a standard sample - large crystallites free of defects, very often a silicon powder -

$$\beta_i^2 = \beta_{calculated}^2 - \beta_{instrument}^2 \tag{7}$$

The usual range of sizes which can be studied by LBA is 2-100 nm. It is important to note that LBA gives crystallite size which is smaller than the particle size in the case of polycrystalline particles. The complete line profile analysis, which will not discussed here, gives, not only a mean particle size, but also the distribution of the particle size. But the Scherrer formula properly applied (using the integral breadth) is often adequate especially in the case of narrow size distribution.

2. 2 Small-Angle X-Ray Scattering (SAXS)

Small-Angle X-Ray Scattering (SAXS), although it is an accurate technique for measuring particle size and specific surface area, is very little used because SAXS requires both dedicated equipment and skilled scientists for the data analysis. A beam of X-rays can be scattered by small domains distributed inside a continuous medium when these heterogeneities (particles, voids) have an electron density different from the carrier (support). The scattered beam gets broader as the domain size decreases.

The SAXS curve is the plot of the scattered intensity $I(s)$ as a function of the scattering parameter $s = (2sin\theta) / \lambda \approx 2\theta / \lambda$ where 2θ is the scattering angle and λ the wavelength. With regard to metal-supported catalysts, the magnitude of the intensity scattered by metallic particles is very often higher than those scattered by the support and the pores of the support because of the high electron density of the metal.

Guinier has shown that, for identical particles, the volume of the particle and its radius of gyration R_G - defined as the mean distance from the electrons in the particle to its center of gravity - can be derived from the central part of the SAXS curve. When the particles are polydispersed, R_G is an average radius in which the larger particles have more weight.

From the high angle part of the SAXS curve, the Porod law yields the specific surface area which can be directly compared to that measured by chemisorption. Assuming a spherical form, we obtain the Porod radius which corresponds to a volume-area mean radius $\Sigma n_i r_i^3 / \Sigma n_i r_i^2$.

When the particles are assumed to be all of equal shape (i.e. spherical) but differ in size, the analysis of the whole profile of the SAXS curve gives volume and surface-weighted distribution of particle sizes using numerical analysis methods or the Fourier transform of the scattered intensity.

Finally, it should be noted two advantages of the SAXS: (i) it is a sensitive technique: for instance supported catalysts containing 0.5 wt % of platinum can be studied; (ii) the size of particles constituted by several crystallites, as well as the particle sizes of poorly crystalline or even amorphous solids, can still be measured by SAXS.

2. 3 X-Ray Absorption Spectroscopy: EXAFS and XANES

Extended X-ray absorption fine structure (EXAFS) and related X-ray absorption techniques (X-ray absorption near-edge structure (XANES)) are described in several books [15, 16] and many review articles and conferences proceedings have been published. A survey of the application of EXAFS to the catalyst characterization was given by B. Moraweck [17].

2. 3. 1 Theoretical Aspects

During the last twenty years, EXAFS has proved to be a powerful tool to characterize catalytic materials. An absorption of X-ray is observed when the energy of the X-rays incident beam E is enough to eject an electron from a core level of an atom of the sample. This ejected photoelectron is backscattered by the surrounding atoms. The photoelectron can be represented as a wave. There is interference - constructive or destructive - between outgoing wave and backscattered waves of the electron, and the total amplitude of the electron wave is enhanced or reduced depending on the energy of the X-rays beam, the distance between the atoms, and the phase shifts due to the backscattering. As a result, the absorption coefficient μ shows oscillations which constitute the XANES and EXAFS signals (Figure 6).

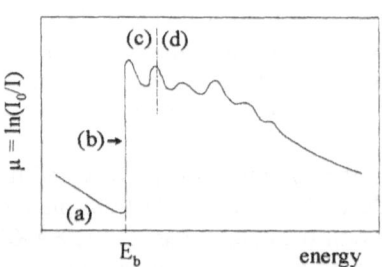

Figure 6: Typical absorption spectrum $\mu = ln(I_0/I)$ as a function of incident X-ray energy; I_0: incident X-ray intensity, I: transmitted X-ray intensity, E_b: binding energy of ejected electron, (a): pre-edge region, (b): edge (or threshold), (c): XANES region, (d): EXAFS region.

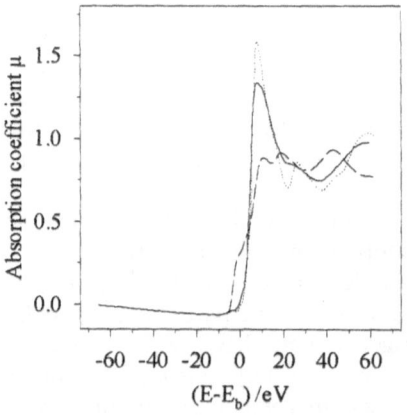

Figure 7: X-Ray absorption edge structure of metallic nickel (dashed line), nickel oxide (dotted line), and Pd-Ni/SiO$_2$ sample (solid line) [14].

The EXAFS region extends from *ca.* 50 eV to 1000 eV above the edge. The X-ray absorption near-edge structure (XANES) covers the range between the threshold and the point at which EXAFS begins.

Local structure information is extracted on the scale of interatomic distances from EXAFS measurements. A considerable advantage is that we know the kind of atom around which the local order is determined in the case of polyatomic sample. The radial distribution function obtained around the absorbing atom gives distance between the absorbing atom and its surrounding neighbours, the number and the nature of the neighbours. XANES gives information on the atomic geometrical arrangement around the absorbing atom and thus to its valence. XANES is also a probing tool to study the electronic properties.

2. 3. 2 Application of EXAFS to the Study of Metal Catalysts

A large number of studies have been performed on metallic supported catalysts. EXAFS was extensively used to follow the successive steps of the catalysts preparation (impregnation, drying, calcination, reduction, activation) depending on the metal loading, the preparation method, the nature of the support and the precursor. A good example is the extensive work on Pt/Al_2O_3 performed at LURE [18]. For instance, it was found that, after reduction at 400 °C, the coordination number of platinum was 5.4 for a 0.7 wt. % Pt content. Such low coordination number means that the platinum particles are composed of 10-15 atoms. Deactivation and regeneration of catalysts were also often studied by EXAFS.

Modern metallic catalysts for selective hydrogenation or naphta reforming are now composed of two or more metals. Because of the atom specificity of EXAFS, the relative organization of the two metal atoms in the bimetallic phase can be determined.. The nice study of Sinfelt et al. [19] on bimetallic $Ru-Cu/SiO_2$, is a well-known example which can be found in several reviews [20, 21] and textbooks [22]. Cu and Ru K-edge EXAFS data of Ru/SiO_2, Cu/SiO_2 and $Ru-Cu/SiO_2$ catalysts after reduction and exposure to oxygen clearly demonstrated that the Ru-Cu bimetallic particles consisted of a ruthenium core and an outer shell of copper atoms.

2. 3. 3 Application of XANES to the Study of Metal Catalysts

At the present time, XANES does not yet reach the same level of development as EXAFS and it is often difficult to obtain quantitative results. Nevertheless, interesting qualitative conclusions can be drawn from XANES spectra. For example, Faudon et al. [14] have shown that the nickel K edge structure of their exchanged $Pd-Ni/SiO_2$ catalyst was quite different from that of the reference Ni foil and close that of nickel oxide, despite of a relatively high loading of Pd (71 Pd atom %) and a reduction temperature of 840 K (Fig. 7). The conclusion was that an appreciable proportion of nickel was in interaction with oxygen atoms of the silica and that the presence of nickel silicate was not impossible. Another example is the study of X-ray absorption edges in platinum-based alloys by Hlil et al [23]. The simultaneous change on the Pt L_{III} and L_{II} edges and on the 3d metal K edge (Fe, Co, Ni) was explained by an electron transfer to platinum from the 3d metal and a population rearrangement of the 3d transition metal states.

2. 3. 4 Conclusions

To sum up, X-ray absorption is a powerful analytical tool to study the structure of catalysts on the atomic scale. It is a precious help to optimize the preparation of catalysts. Kinetics studies under controlled atmosphere, temperature, and pressure are now possible owing to the availability of quick or dispersive EXAFS. But it should be noted that the analysis of EXAFS data requires expertise to obtain unambiguous results.

3. ELECTRON MICROSCOPY

Electron microscopy is very popular in the field of catalyst characterization because this technique allows the direct observation of the catalyst morphology. Several reviews have been devoted to its application to catalyst characterization [5, 24-27].

Electron microscopes can be divided into two categories: the conventional transmission electron microscopes (CTEM or TEM) and the scanning electron microscopes. In the CTEM, the image is obtained using a large and parallel electron beam. In the second category, a fine electron beam is scanned over the specimen and the magnification depends on the scanned area. Electron microscopes are often equipped with attachments giving analytical capabilities: nanodiffraction for phase identification on nanodomains, energy dispersive X-ray emission spectroscopy (EDX) for general elementary analysis (e.g. determination of the composition of bimetallic particles) and electron energy loss spectroscopy (EELS) for special elementary analysis and structural information.

3. 1 Preparation of Samples

Specimen preparation is a critical step in electron microscopy because the quality of the results are highly dependent on the thickness of the sample which should be less than 0.05 to 0.1 micron. If the catalyst grains are smaller than 0.05-0.1 µm, a direct observation is possible after dispersion of the powder on a grid. When the grains are too thick, they can be cut into thinner sections by ultra microtomy after embedding in a polymeric resin. This type of preparation enables imaging of the internal morphology and metal dispersion in micro- or mesoporous supports. To visualize particles on the surface of supports too thick or containing heavy elements (rare-earth oxides), extractive replicas have to be made: the dispersed catalyst powder covered by a sputtered carbon film is plunged into an acid solution to dissolve the catalyst support without dissolving the metal particles.

3. 2 The Different Types of Images and Their Use for the Study of Metal
Catalysts

Several types of images can be obtained. In the transmission microscope, the parallel electron beam is partially scattered by the specimen and the normal image (bright-field image) is formed by the combination of the scattered and unscattered electrons: the particle appears darker than the background. Detection of particles or phases in the specimen is limited primarily by contrasts effects. The contrast depends upon the thickness, the atomic number and the crystallinity of the phases of the specimen; the finest image will be obtained on high-Z, well-crystalline metallic particles supported on thin, low-Z, amorphous supports.

From the electron microscopy views, the size, the shape and the distribution of metal particles on the support can be characterized. Particles can be homogeneously dispersed on/in the support or concentrated in preferential place (e.g. along the defects of the support or outside the microporous lattice of the zeolite). Electron microscopy is widely used to determine the size distribution and the average size of metal particles. The size distribution can be established by measuring about 1000 particles on different zones of several views. Several types of distribution and average can be calculated as discussed in Section 1.1. Note that if a rough examination of views shows that the sizes are too dispersed, a granulometry is probably useless because relations between morphology and catalytic properties cannot be unambiguously established.

Another type of image can be obtained for crystalline specimen by collecting only one diffracted beam, excluding all the unscattered electrons: the image of the particle is bright against a dark background (dark-field image). Only the particles in Bragg orientation will be detected. This technique allows to obtain a three-dimensional description of the particle morphology. In a multiphase specimen, a specific phase can be imaged.

By collecting all the diffracted beam from a selected area (1-1000 nm) in the back focal plane (i.e. before the image formation), a micro- or nanodiffraction pattern is obtained. From the position of the spots, phase identification is possible, but the precise determination of the lattice parameters is difficult. However, information on the preferential orientation of the particles on the support and on the nature of the exposed crystallographic planes of the particles can be obtained. Epitaxial relation between metal particle and support were demonstrated by nanodiffraction. Thus it was shown by Gallezot *et al.* [28] that 2-nm platinum particles are not oriented on charcoal but epitaxial relation takes place on graphite: the fcc lattice of Pt particles are distorted so that the platinum atoms tend to coincide with the carbon atoms of the basal planes of graphite; the smaller the particle, the larger the distortion.

3. 3 Energy Dispersive X-Ray Emission Spectroscopy (EDX)

The composition of heterogeneous catalysts can be determined by energy dispersive X-ray emission spectroscopy (EDX). Electrons of the incident beam can inelastically interact with the specimen to eject core electrons (K, L, M...). The de-excitation of the atoms occurs by emission of characteristic X-ray lines. Thus elementary analysis, both qualitative and quantitative, is possible provided that the atomic number of the element is higher than 5. The analyzed zone of the studied specimen can be varied from several hundreds of nanometers to verify the homogeneity or to map an element to one namoneter (local analysis) depending on the microscope type.

For bimetallic catalysts, it is important to know the composition of individual metal particles (proportion of the two metals) to establish the distribution of composition of particles [14]. This is possible on particles as small as 1-1.5 nm, and for particles larger than 5 nm a composition profile can even be determined in a given particle to detect segregation [27]. Note that no chemical information (*e.g.* valence) can be obtained by EDX. A point has to be considered. The composition of the specimen can be modified under the electron beam, especially the small particles: cation reduction followed by particle formation, sintering, vaporization of the most volatile metal... In addition to these limitations, the signal of the small particles is low. Therefore the measured composition of the smallest particles can be inaccurate.

3. 4 Electron Energy Loss Spectroscopy (EELS)

Electron energy loss spectroscopy (EELS), although it is not so used as EDX, can provide   not only analytical information but also structural information. The energy loss of the electrons of the incident beam interacting with the specimen is measured. The EELS spectrum - number of collected electrons as a function of the energy loss - shows peaks and structures known as extended energy loss fine structure (EXELFS) and energy loss near edge structure (ELNES) similar to EXAFS end XANES, respectively. The EELS technique is useful for quantitative analysis of light elements (C, O) not or badly detected by EDX. Structural information obtained by EELS (local environment, valence state) is only qualitative because of the low signal-to-noise ratio of the data. Thus, Gallezot et al. [29] have studied the coke deposits on Pt/Al_2O_3 by comparing the EELS spectra of the catalysts with those of reference compounds. From the EELS spectra of Ni on CeO_2 microcrystals recorded across several metal-oxide interfaces, Colliex et al. [30] has shown that two types of contacts can be discriminated: "clean" ones with metallic Ni on CeO_2 and "contaminated" ones with oxidized Ni on reduced CeO_{2-x} supports.

3. 5 Conclusions

To sum up, electron microscopy, because it allows a direct, local observation, has now become a routine technique for catalyst characterization. But the following points should be not forgotten. A very small quantity of matter is observed on a micrograph and care must be taken in the generalization to the whole catalyst. The use of vacuum and an electron beam can modify the catalyst. It is very difficult to discriminate between elements having close atomic numbers or to study catalysts based on a support having a high atomic number.

References

[1] van Hardeveld R. and Hartog F., Surf. Sci. 15 (1969) 189-230.

[2] Mackay A.L., Acta Cryst. 15 (1962) 916-918.

[3] Gordon M.B., Ph. D. Thesis (3ème Cycle), Grenoble, France, 1978.

[4] Rousset J.L., Cadrot A.M., Cadete Santos Aires F.J., Renouprez A., Mélinon P., Perez A., Pellarin M., Vialle J.L. and Broyer M., J. Chem. Phys. 102 (1995) 8574-8585.

[5] Bergeret G. and Gallezot P., Particles Size and Dispersion Measurements, in "Handbook of Heterogeneous Catalysis", G. Ertl, H. Knözinger and J. Weitkamp Eds. (VCH, Weinheim, to be published May 1997) Part A, Section 3.1.2.

[6] Klug H.P. and Alexander L.E., X-Ray Diffraction Procedures for Polycrystalline and Amorphous Materials, 2nd ed. (Wiley, New York, 1974).

[7] Modern Powder Diffraction. D.L. Bish and J.E. Post Eds., *Reviews in Mineralogy*, Vol. 20 (Mineralogical Society of America, Washington DC, 1989).

[8] Gallezot P, X-Ray Techniques in Catalysis, in "Catalysis, Science and Technology". J.R. Anderson and M. Boudart Eds. (Springer, Berlin. 1984) Vol. 5, pp. 221-273.

[9] Bergeret G., X-Ray Powder Diffraction, in "Handbook of Heterogeneous Catalysis", G. Ertl, H. Knözinger and J. Weitkamp Eds. (VCH, Weinheim, to be published May 1997) Part A, Section 3.1.3.1.

[10] Powder Diffraction File (PDF). International Centre for Diffraction Data. 12 Campus Blvd, Newton Square, PA 19073-3273 USA, http://www.icdd.com.

[11] Caussin P., Nusinovici J. and Beard D.W., in "Advances in X-Ray Analysis", C.S. Barrett. J.V. Gilfrich, R. Jenkins, T.C. Huang and P.K. Predecki Eds., (Plenum, New York, 1989) Vol. 32, pp. 531-538; Diffrac-AT, Socabim. 9 bis Villa du Bel-Air, F75012 Paris.

[12] Hardacre C., Rayment T. and Lambert R.M., *J. Catal.* **158** (1996) 102-108.

[13] Clausen B.S., Steffensen G., Fabius B., Villadsen J., Feidenhans L.R. and Topsoe H., *J. Catal.* **132** (1991) 524-535.

[14] Faudon J.F., Senocq F., Bergeret G., Moraweck B., Clugnet G., Nicot C. and Renouprez A., *J. Catal.* **144** (1993) 460-471.

[15] Koningsberger D.C. and Prins R., X-Ray Absorption. Principles, Applications. Techniques of EXAFS, SEXAFS, and XANES (Wiley, New York, 1988).

[16] Teo B.K., EXAFS: Basic Principles and Data Analysis (Springer, Berlin, 1986).

[17] Moraweck B., X-Ray Absorption Spectroscopy: EXAFS and XANES, in "Catalyst Characterization: Physical Techniques for Solid Materials". B. Imelik and J.C. Védrine Eds. (Plenum, New York, 1994) pp. 377-416.

[18] Lagarde P, Murata F., Vlaic G., Freund E., Dexpert H. and Bournonville P., *J. Catal.* **84** (1983) 333-343; Berdala J., Freund E. and Lynch J.P., *J. Phys. Fr. Coll C8* **272** (1986) 265-268 and 269-272; Bazin D., Dexpert D., Lagarde P. and Bournonville P., *J. Catal.* **110** (1988) 209-215.

[19] Sinfelt J.H., Via G.H. and Lytle F.W., *J. Chem. Phys.* **72** (1980) 4832-4844.

[20] Sinfelt J.H., Via G.H. and Lytle F.W., Applications of EXAFS in Catalysis. Structure of Bimetallic Cluster catalysts, *Catal. Rev.-Sci. Eng.* **26** (1984) 81-140.

[21] Prins R. and Koningsberger D.C., in ref [15], pp. 358-363.

[22] Niemantsverdriet J.W., Spectroscopy in Catalysis: an Introduction (VCH, Weinheim, 1995), pp. 150-164.

[23] Hlil E.K., Baudoing-Savois R., Moraweck B. and Renouprez A.J., *J. Phys. Chem.* **100** (1996) 3102-3107.

[24] Howie A., The Study of Supported Catalysts by Transmission Electron Microscopy, in "Characterizationj of Catalysts", J.M. Thomas and R.M. Lambert Eds. (Wiley, Chichester, 1980) pp. 89-104.

[25] Baird T., Characterization of Catalysts by Electron Microscopy. in " A Specialist Periodical Report" Catalysis. Vol. 5 (Royal Society of Chemistry. London. 1982) pp.172-219.

[26] Sanders J.V., The Electron Microscopy of Catalysts, in "Catalysis, Science and Technology". J.R. Anderson and M. Boudart Eds.. Vol. 7 (Springer. Berlin, 1985) pp. 51-158.

[27] Gallezot P. and Leclercq C., Characterization of Catalysts by Conventional and Analytical Electron Microscopy. in "Catalyst Characterization: Physical Techniques for Solid Materials". B. Imelik and J.C. Védrine Eds. (Plenum. New York, 1994) pp. 509-558.

[28] Gallezot P. Leclercq C., Mutin I., Nicot C. and Richard D., *J. Microsc. Spectrosc. Electron.* **10** (1985) 479-484.

[29] Gallezot P. Leclercq C., Barbier J. and Marecot P., *J. Catal.* **116** (1989) 164-170.

[30] Colliex C., Lefèvre E. and Tencé in "Electron Microscopy and Analysis 1993, Proc. of the Institute of Physics. Electron Microscopy and Analysis Group Conference. Liverpool, 1993". A.J. Craven Ed.. Institute of Physics Conference Series Number 138. Institute of Physics Publishing. Bristol 1993. pp. 25-30.

Vibrational Spectroscopy with Neutrons

H. Jobic

*Institut de Recherches sur la Catalyse, CNRS, 2 avenue Albert Einstein,
69626 Villeurbanne, France*

1. INTRODUCTION

Several neutron techniques are being used to study catalytic systems [1]. We will
limit ourselves here to inelastic neutron scattering (INS) which is a method of great
potential interest to study vibrational modes of catalysts and of adsorbed molecules.
INS is one of the numerous vibrational techniques available for a better
understanding of surface phenomena. Each technique has its particular advantages
for a given system in terms of spectral domain, resolution, sensitivity and
experimental conditions. For example infrared spectroscopy is a very efficient
method to detect adsorbed CO (see the contribution by Maugé et al.) but it is much
less sensitive to adsorbed hydrogen.

Because of its penetration power, neutron is not a priori a surface probe.
However it is possible to obtain spectra of surface species provided that the
adsorbate has a larger neutron cross section than the substrate. Since the hydrogen
atom has the largest neutron scattering cross section, it is the preferred probe for
adsorption studies. Nevertheless, one must put into the beam about a one hundredth
of a mole of hydrogenous compound to have a reasonable signal. Therefore, only
high surface area catalysts can be studied. The consequence is that real catalysts,
supported or unsupported, can be tackled by neutron scattering; experiments with
single crystals cannot be considered today because the quantity of adsorbed species
is too small.

The applications of INS to catalysis have been mainly focused to systems
which are either difficult or impossible to study by other spectroscopies such as
transmission or reflection-absorption infrared, and Raman. This happens because
the sample is usually opaque or completely black so that it may have only limited

frequency windows in infrared or it may decompose or fluoresce in the laser beam in Raman. Comparisons have been made in a few cases with the results obtained by electron energy loss spectroscopy (EELS).

The kind of catalyst which is studied in INS has generally an inhomogeneous surface, e. g. oxides, sulfides and metals, although zeolites. which are well-crystallised materials, are well suited. These substrates can be almost transparent to neutrons if they contain a small quantity of hydrogen. in which case the neutron spectrum will be fairly flat and it will be possible to observe all the vibrational modes of the adsorbate.

2. THEORY

2.1 Interaction of neutrons with matter

The neutron wavelength is given by the de Broglie relation $\lambda = h/mv$. where h is Plank's constant and v the neutron velocity. The associated wave vector $\mathbf{k}$ has the magnitude $k = 2\pi/\lambda$. This allows to define the neutron momentum as $\mathbf{p} = mv = \hbar\,\mathbf{k}$. and the neutron energy as $E = \frac{1}{2}mv^2 = \hbar^2 k^2 / 2m$.

A major difference with other spectroscopic techniques is that neutrons interact only with the atomic nuclei, whereas photons for example interact with the electronic cloud around the nuclei. The interaction of neutrons with matter is thus relatively weak.

It is convenient to separate the total scattering cross section into two parts: a coherent and incoherent one. Coherent scattering takes into account interference effects between the waves scattered from each nucleus. Incoherent scattering corresponds to the mean-square deviation from the mean potential. Deviations from the average are due to isotopic or spin effects. Coherent and incoherent cross sections are given in Table I for some elements in barns (1 barn=10^{-28} m^2). Neutron absorption is small for most nuclei. except for a few notable exceptions like boron and cadmium.

It appears that, unlike with X-rays, the scattering cross sections vary irregularly from one atom to another, even from one isotope to another.

In the case of diffraction experiments, large incoherent scattering should be avoided since it increases the background below the Bragg peaks. It is thus advised to use deuterated compounds for these experiments.

For inelastic studies, the energy of the scattered neutrons is analysed and both coherent and incoherent scattering can be measured. For single crystals. coherent scattering allows a determination of the phonon dispersion curves. For surface work a high surface to bulk ratio is required for the substrate, and the adsorbates must have enhanced cross sections (i.e. contain H) to obtain the best contrast.

Table I: Coherent, incoherent and absorption cross sections for some elements (in barns, 1 barn=10^{-28} m^2, σ_{abs} is proportional to λ, here $\lambda = 1$ Å).

Element	Atomic number	Mass number	σ_{coh}	σ_{inc}	σ_{abs}
H	1		1.7586	79.9	0.19
		1	1.7599	79.91	0.19
		2 = D	5.597	2.04	0
		3 = T	3.07	0	0
B	5		3.54	1.7	426.
C	6		5.554	0.001	0
		12	5.563	0	0
		13	4.81	0.034	0
N	7		11.01	0.49	1.1
O	8		4.235	0	0
Al	13	27	1.495	0.009	0.13
Si	14		2.163	0.015	0.1
S	16		1.019	0.007	0.3
Ni	28		13.3	5.2	2.5
Zr	40		6.44	0.16	0.1
Ru	44		6.53	0.07	1.42
Cd	48		3.3	2.4	1400.
Pt	78		11.65	0.13	5.72

2.2 Vibrational Spectroscopy with neutrons

The two basic quantities in a neutron scattering experiment are the energy transfer $\hbar\omega$ and the momentum transfer $\hbar \mathbf{Q}$ which are defined as

$$\hbar\omega = E_0 - E' = (\hbar^2 / 2m)(k_0^2 - k'^2) \qquad (1)$$

$$\hbar\mathbf{Q} = \hbar(\mathbf{k}_0 - \mathbf{k}') \qquad (2)$$

where $\mathbf{k}_0$ and $\mathbf{k}'$ are, respectively, the incident and scattered wave vectors. Elastic scattering corresponds to $k' = k_0$, so that only momentum is transferred ($\hbar\omega = 0$). When the neutron exchange also energy with the sample, $\hbar\omega$ will be positive for k' $< k_0$ (energy loss or creation of an excitation) and the scattering is called inelastic.

In INS experiments, the measured intensity is proportional to the partial differential cross section d$^2\sigma$/dΩdE, which represents the number of neutrons scattered into a solid angle dΩ with energy in the range dE. As a first approximation only the incoherent part can be considered if the sample contains hydrogen atoms. It can be calculated for a harmonic oscillator and for a polyatomic molecule by resolving molecular vibrations into normal modes. For a sample at low temperature, a relatively simple expression can be obtained for a fundamental λ in neutron energy loss :

$$\frac{d^2\sigma}{d\Omega dE} = \frac{k'}{k_0}\frac{\sigma_{inc}}{4\pi}\sum_d \exp(-Q^2 <u^2>)\frac{\hbar\left|Q.C_d^\lambda\right|^2}{2m_d\,\omega_\lambda}\,\delta(\omega - \omega_\lambda) \qquad (3)$$

The intensity of the δ function at frequency ω_λ, corresponding to the normal mode λ, is governed by the product $\left|Q.C\right|^2$. The vector C_d^λ describes the displacement of the dth atom during the λth normal mode, in mass-weighted Cartesian coordinates. These vectors can be obtained from empirical force fields or from ab initio quantum chemical methods. The Debye-Waller factor, $\exp(-Q^2<u^2>)$, reflects the dynamical disorder, $<u^2>$ being the mean-square displacement due to the normal modes. This factor decreases the intensity as the momentum transfer Q increases (on most spectrometers, k' is small so that the energy transfer is proportional to Q^2, Eqs. 1 and 2).

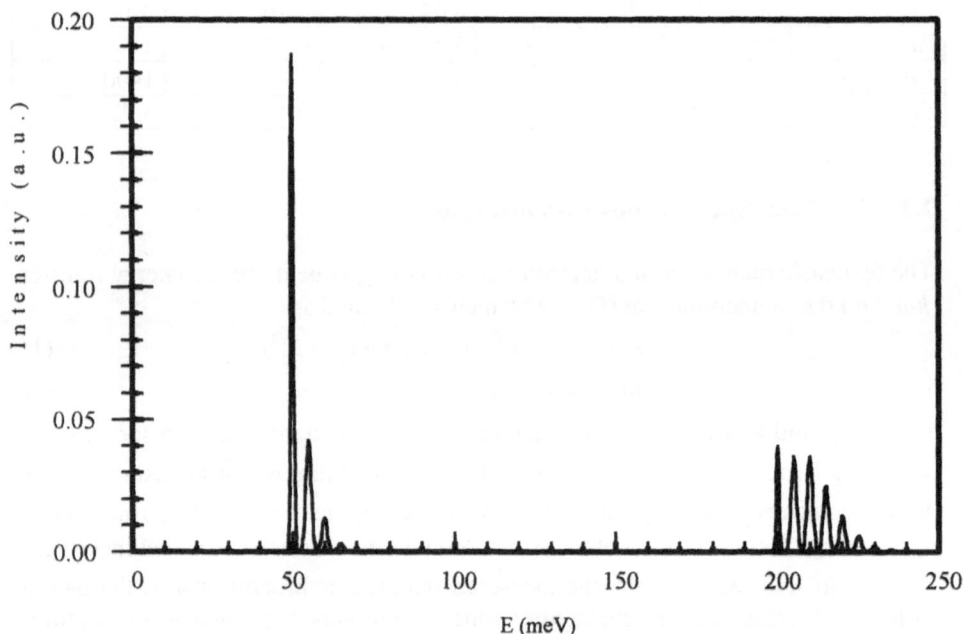

Figure 1 : Simulation of the effect of the Debye-Waller factor on the redistribution of intensity for two modes at 50 and 200 meV (1meV=8.065 cm^{-1}).

In reality, the spectrum is complex because overtones and combinations are present, however their intensities can also be calculated. A further complication occurs through the convolution of the high-frequency internal modes by the low-frequency external modes. The scattering function corresponding to the external modes can be expanded in terms of one-phonon, two-phonon..., with another Debye-Waller factor due to the external modes. This factor further decreases the intensity at the frequency ω_λ. The intensity which is taken from the fundamental is redistributed into side bands. An example is shown in Figure 1 for two modes initially of equal intensity, at 50 meV ($\approx$ 400 cm^{-1}) and 200 meV ($\approx$ 1600 cm^{-1}). After taking into account the effect of the Debye-Waller factor, the intensities are greatly modified. At small energy transfer, a large fraction of the intensity is contained in the fundamental, but at high energy transfer, the total intensity from the side bands is larger than the zero line. This is due to the fact that Q^2 is 4 times larger at 200 meV compared to 50 meV. For a larger value of $<u^2>$, the intensity of the zero line can even become negligibly small.

The complete calculation of an INS spectrum may look complicated. However analysis packages are now available to simulate INS spectra or to refine a force field directly to the observed profile.

A good overview of spectroscopic applications of INS is contained in a special issue of Spectrochimica Acta [2]. The main INS characteristics can be summarised as such :

⇨ Neutrons are very sensitive to hydrogen because of its large incoherent cross section, σ_H, and its low mass, m_H. However the scattering from other atoms can be measured.

⇨ There are no considerations on the symmetry of the vibrations so that all modes can be observed. This is in clear contrast with the selection rules operating in infrared and Raman spectroscopies.

⇨ The modes involving large displacement vectors, $\mathbf{C}^\lambda$, of light atoms will have high intensities. For a powder sample, an average over all possible orientations of $\mathbf{C}$ relative to $\mathbf{Q}$ must be performed.

⇨ Information on the vibrational modes can be obtained from the frequencies and from the displacement vectors. INS is therefore the only vibrational technique where spectra can be calculated with reasonable accuracy.

3. EXPERIMENTAL

Neutrons can be produced from steady-state reactors (e.g. the high flux reactor at the Institut Laue-Langevin, Grenoble, France), and also from pulsed sources (e.g. the spallation source ISIS at the Rutherford Appleton Laboratory, Didcot, UK). The source brightness is however several orders of magnitude lower than X-ray synchrotrons. Neutrons issued from the source are slowed down in moderators whose temperatures can vary between 25 and 2000 K. This gives neutron wavelengths ranging from 0.2 to 20 Å. The beam size at sample position is typically 5×2 cm^2 and large sample environments can be accommodated around the sample: cryostat, furnace, high-pressure cell, etc.

The number and diversity of neutron instruments around a neutron source is large. Each instrument is designed to cover a given range of momentum transfer and/or time scale. There are two main classes of instruments: elastic scattering instruments which are used for determining the structure of materials, and inelastic scattering spectrometers which, by measuring energy transfers, give information on atomic and molecular motions. For INS, the whole spectral range 1-4000 cm^{-1} can now be observed with present neutron spectrometers and the energy resolution is reasonable (typically, 5 cm^{-1} < ΔE < 60 cm^{-1} or ΔE/E $\approx$ 2%).

In order to define the quantity of sample required for an INS experiment, the following relation can be used

$$I \,/\, I_0 = \exp(-\sum_i N\sigma_i \, \ell) \qquad\qquad (4)$$

where $I \,/\, I_0$ gives the transmission of the sample perpendicular to beam. N is the number of scatterers per unit volume, σ_i their total cross section and ℓ the thickness of the sample.

The percentage of scattering is usually aimed to be 10% , to avoid multiple scattering. If we consider for example liquid benzene, its density is 0.8765 g/cm^3, its molecular weight 78.12 g/mol and one finds using Table 1 : $\Sigma \sigma_i$ = 529 $\times$ 10^{-28} m^2 (for λ = 5 Å). For a flat sample 0.03 cm thick, this gives a transmission of 90%. The beam area at the sample position being typically of 10 cm^2, it appears that a quantity of 0.26 g of benzene is sufficient.

However, in order to put into the beam the same number of protons for catalytic studies, the quantity of catalyst results to be much larger. If we consider for example hydrogen dissociated on an unsupported metal powder of 50 m^2/g of surface area. The average number of hydrogen atoms adsorbed per square meter will be of the order of 10^{19} at saturation. In order to obtain 10% scattering, one thus needs 20 g of catalyst. If smaller coverages have to be studied, the quantity of powder should be augmented.

The substrates which can be studied by INS are unsupported metals (Raney or colloidal black), or small metallic particles supported on silica or alumina. Different oxides have also been studied, as well as sulfides or charcoal. Zeolites offer very high surface areas. All these powders are widely used in catalysis and cannot be easily studied by other vibrational techniques.

Special care must be taken to avoid pollution by water or hydrocarbons after preparation. The neutron cells can be cylindrical or slab-shaped, depending on the neutron beam size and scattering strength. They are usually made from aluminium, but stainless-steel or quartz cells have also been used. More recently, containers made of zircalloy, an alloy based on zirconium, have been utilised because this metal has a small absorption (Table 1) and a mechanical resistance comparable to stainless-steel.

For molecular systems, theory and experiment shows the necessity to record INS spectra at very low temperature (around 5 K); this will decrease the relative intensity of multiphonon features and sharpen the fundamentals. When the total Debye-Waller factor is larger, e.g. for an hydrogen atom bonded to a rigid framework (metal or zeolite), measurements can be performed at room temperature.

4. EXAMPLES

As applications of INS to catalysis, one can mention studies concerning hydrogen chemisorbed on metals, sulfides and oxides, hydroxyl groups; organometallic compounds; hydrocarbons and water on different catalysts and zeolitic systems. Some recent examples will now be described.

4.1. Hydrogen chemisorption

Hydrogen chemisorbed on high specific surface area materials is difficult to characterise by optical spectroscopies. EELS data have been obtained on single crystals and since only a limited number of sites are available on well-defined crystal planes, the results have been found to be useful for assigning spectra obtained on polycrystalline materials.

With INS, all the local modes of hydrogen can be observed because there are no selection rules, which is clearly an advantage. However, if there are different planes exposed on the surface or if there are several species, the assignment can be more difficult. In the case of hydrogen bonded to heavy atoms, the heavy atoms can be considered as fixed during the local modes of hydrogen. A consequence is that the mean-square amplitudes for the hydrogen atoms will be generally small so that the sample temperature will be of small influence on the INS intensities (between 5 and 300 K). Another consequence is that the nondegenerate modes of hydrogen will have nearly the same intensity and an E mode will be twice as intense as an A mode. The INS spectra can be easily fitted and the relative intensities of the bands yield the populations of the various sites.

4.1.1 Hydrogen on nickel

If we consider for example hydrogen on nickel powders, all the recent INS studies indicate that hydrogen is predominantly multiply bonded, under (μ_3-H) form [3-5].

However, some authors still claim that after dissociation one hydrogen atom is only bonded to a single nickel atom (e. g. ref. 6), in complete disagreement with theoretical and experimental results, including the work on single crystals.

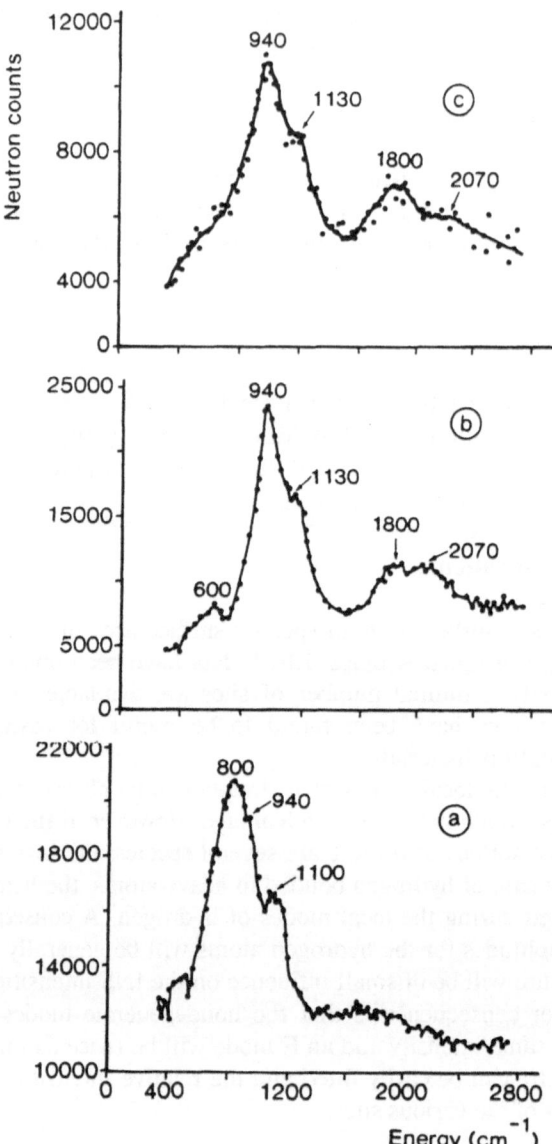

Figure 2 : INS spectra of hydrogen adsorbed on a doped Raney nickel :

  (a) residual hydrogen, after desorption at 373 K.

  (b) spectrum obtained after adsorption under 13 mbar of H_2 (the spectrum of residual hydrogen has been subtracted).

  (c) spectrum obtained after adsorption of H_2 at atmospheric pressure (spectra a and b have been subtracted).

All the spectra were recorded at 300 K with the INFB spectrometer at the ILL.

In the first INS studies on Raney nickel [3-4], the catalyst was outgassed at a sample temperature of 573 K, and it was clear from INS that no hydrogen was left on the surface. However, sintering occurs in these conditions and the surface area was found to decrease from 130 to 40 m^2g^{-1}.

The spectrum shown in Figure 2(a) was obtained after outgassing at only 373 K [5]. The signal produced by the aluminium container and the phonons due to the nickel sample have been subtracted. INS intensities show that a large quantity of hydrogen is present on this catalyst, about 30% of the amount originally adsorbed (this quantity depends on the pumping rate). At this temperature, no structural modification is induced but water has been completely removed. The spectrum of residual hydrogen consists of two bands at 800 and 1100 cm^{-1} sitting on a large background. The background is due to a broad distribution of sites for hydrogen. Even if the surface is mainly made of low index planes, the surface is far from being perfect : defects, steps, and kinks are known to occur. The two modes at 800 and 1100 cm^{-1} were assigned to the antisymmetric (E) and symmetric (A) stretching modes of hydrogen atoms adsorbed on sites of nearly C_{3v} symmetry, located on the (110) faces. The weaker contribution observed at 940 cm^{-1} was assigned to hydrogen adsorbed on C_{3v} sites, on (111) faces.

The two modes of this species are more clearly observed in Figure 2 (b) which corresponds to the spectrum obtained after adsorption under an equilibrium pressure of 13 mbar of H_2. The contribution from residual hydrogen has been subtracted and it can be noted that the base line is closer to zero. The band at 940 cm^{-1} is assigned to the antisymmetric stretch (E) of hydrogen adsorbed on C_{3v} sites, while the shoulder at 1130 cm^{-1} corresponds to the symmetric (A) stretching mode. The overtones and combination of these two modes produce a broad contribution around 2070 cm^{-1}. The band centred at 1800 cm^{-1} is assigned to the stretching mode of on top hydrogen (H bonded to a single Ni atom). The bending mode of this linear species is expected to fall between 800 and 1150 cm^{-1} and would then be hidden by the more intense features due to multiply bonded hydrogen. The weak contribution at 600 cm^{-1} appears to be too low in energy to be assigned to the bending mode of on top hydrogen. It was tentatively assigned to the symmetric stretching mode of a hydrogen atom bonded in a four-fold site, in agreement with a frequency observed by EELS on the (100) faces.

After hydrogen adsorption under atmospheric pressure, and subtraction of all the previous contributions (spectra a and b), the spectrum shown in Figure 2(c) is obtained. The main species is still the one located in C_{3v} sites, but the relative intensity of the band at 1800 cm^{-1} has increased.

In conclusion, after evacuation of the Raney nickel at 373 K, 30% of the total hydrogen uptake remains on the surface. This residual hydrogen is adsorbed on (110) and (111) faces, and also on defects. On top hydrogen is only detected upon readsorption of hydrogen, at higher pressure. This shows that multiply bonded hydrogen is more strongly adsorbed than linear hydrogen. This species appears close to the saturation of the surface. Even if the proportion of linear hydrogen is small (about 15%), it could be the active species in hydrogenation reactions since it has been found from kinetic studies and pulse experiments that only weakly adsorbed hydrogen is active for hydrogenation of acetonitrile [5].

4.1.2 Hydrogen on ruthenium sulfide

On sulfides, it has been postulated that hydrogen may be adsorbed either homolytically or heterolytically. Accordingly, hydrogen chemisorption should form not only SH groups but also metal-hydrogen bonds. Hydroxyl and sulfhydryl groups have previously been measured but hydride-type hydrogen was never clearly evidenced. In particular MoH species have not been observed on MoS_2 in spite of several INS experiments.

We have recently studied by INS the interaction of hydrogen with RuS_2 [7] because this sulfide is ≈ 10 times more active than MoS_2 in hydrogenation and hydrodesulfurization reactions. This could be due to the larger adsorption capacity of ruthenium sulfide or to the presence of different hydrogen species. In order to minimise the background, we have used unsupported RuS_2 consisting of homodispersed spheres with a diameter 45-50 Å. The INS spectra were recorded with the spectrometer TFXA at ISIS (Rutherford Appleton Laboratory, UK). Several hydrogen species have been observed using INS, by changing the degree of reduction and the experimental conditions:

(i) When the catalyst is sulfided at 673 K under H_2S flow. chemical analysis gives a stoichiometry S/Ru = 2.25. The INS spectrum shown in Figure 3(a) indicates only the presence of SH groups on the surface. The peak measured at 737 cm^{-1} is assigned to SH bending modes, in agreement with previous INS studies performed on other sulfides: these modes were measured at 694 cm^{-1} on WS_2 [8] and at 650 cm^{-1} on MoS_2 [9].

(ii) If the catalyst is partially desulfurized under H_2 flow at 513 K. the solid composition becomes $RuS_{1.88}$. After hydrogen adsorption at low pressure (less than 1 mbar), a new peak and a shoulder appear respectively at 823 and 550 cm^{-1}. Figure 3(b). When the hydrogen pressure is increased (0.5 bar at 295 K). the peak at 542 cm^{-1} gains further in intensity, Figure 3(c). This peak and the one measured at 826 cm^{-1} are assigned to the bending modes of two different RuH species. It appears that the hydridic group which gives rise to the peak at 542 cm^{-1} is more weakly adsorbed than the species which yields the peak at 826 cm^{-1}. Since it has been mentioned in the previous section that on Raney nickel reactive hydrogen is the more weakly adsorbed species, it can be proposed that the hydrogen species active on ruthenium sulfide for hydrogenation reactions is the one giving rise to the peak at 542 cm^{-1}.

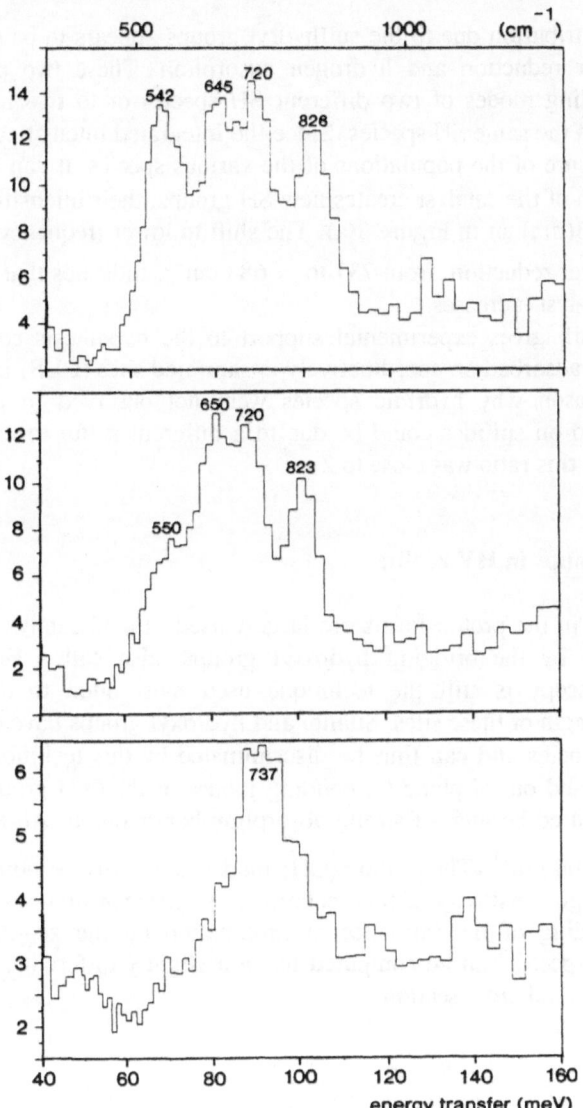

Figure 3 : INS spectra obtained at 25 K with ruthenium sulfide:
  (a) unreduced catalyst.
  (b) hydrogen adsorbed at low pressure (less than 1 mbar) on partially desulfurised RuS_2.
  (c) hydrogen adsorbed at higher pressure (0.5 bar at 295 K) on partially desulfurised RuS_2.

The contribution due to the sulfhydryl groups appears to be split into two components after reduction and hydrogen adsorption. These two peaks can be assigned to bending modes of two different SH species or to two nondegenerate bending modes of the same SH species. Since the integrated intensities of the bands are a direct measure of the populations of the various species, it can be concluded that the reduction of the catalyst creates new SH groups: their intensity is $\approx$ 3 times larger in Figure 3(b) than in Figure 3(a). The shift to lower frequency observed for the SH modes after reduction, from 737 to $\approx$ 683 cm^{-1}, indicates that the Brönsted acidity of the catalyst increases.

This work gives experimental support to the hypothesis concerning the role of hydrogen adsorbed on coordinatively unsaturated sites (CUS) in catalysis by sulfides. The reason why hydridic species were not observed in previous INS studies performed on sulfides could be due to a different sulfur-to-metal ratio (in previous samples this ratio was close to 2).

4.2 Hydroxyl groups in HY zeolite

Zeolites in the proton forms are largely used in acid catalysis. The active sites are formed by the bridging hydroxyl groups, also called Brönsted sites. Infrared spectroscopy is still the technique used most often to determine the location and strength of these sites. Silanol and hydroxyl groups have different O-H stretching frequencies and can thus be discriminated by this technique. However, the in-plane (δ) and out-of-plane (γ) bending modes of the O-H groups cannot be measured by infrared because of strong absorption bands due to Al, SiO$_4$ units in the range 200-1300 cm^{-1}. The δ and γ(OH) modes can easily be observed by INS because of the high sensitivity of this technique to hydrogen motions. To obtain a better understanding of the influence of protonation on the zeolite vibrational modes, the INS spectra can be computed using a slightly different expression for the double-differential cross section:

$$\frac{d^2\sigma}{d\Omega dE} = \frac{k'}{4\pi k_0} \frac{\sigma_d}{m_d} \sum_d \exp(-\mathbf{Q}^2 <\mathbf{u}_d^2>) G_d(\omega) \tag{5}$$

where $G_d(\omega)$ is the atomic amplitude weighted vibrational density of states and σ_d the total scattering cross section for atom d (because of the large momentum transfers implied for energy transfers larger than 400 cm^{-1}, the incoherent approximation can be used). One-phonon processes are predominant for this system because of the relatively small values of the mean-square amplitudes $<\mathbf{u}^2>$ for all atoms, so that only these terms are considered in Eq. (5). This approximation can be checked by the very small temperature dependence of the INS data.

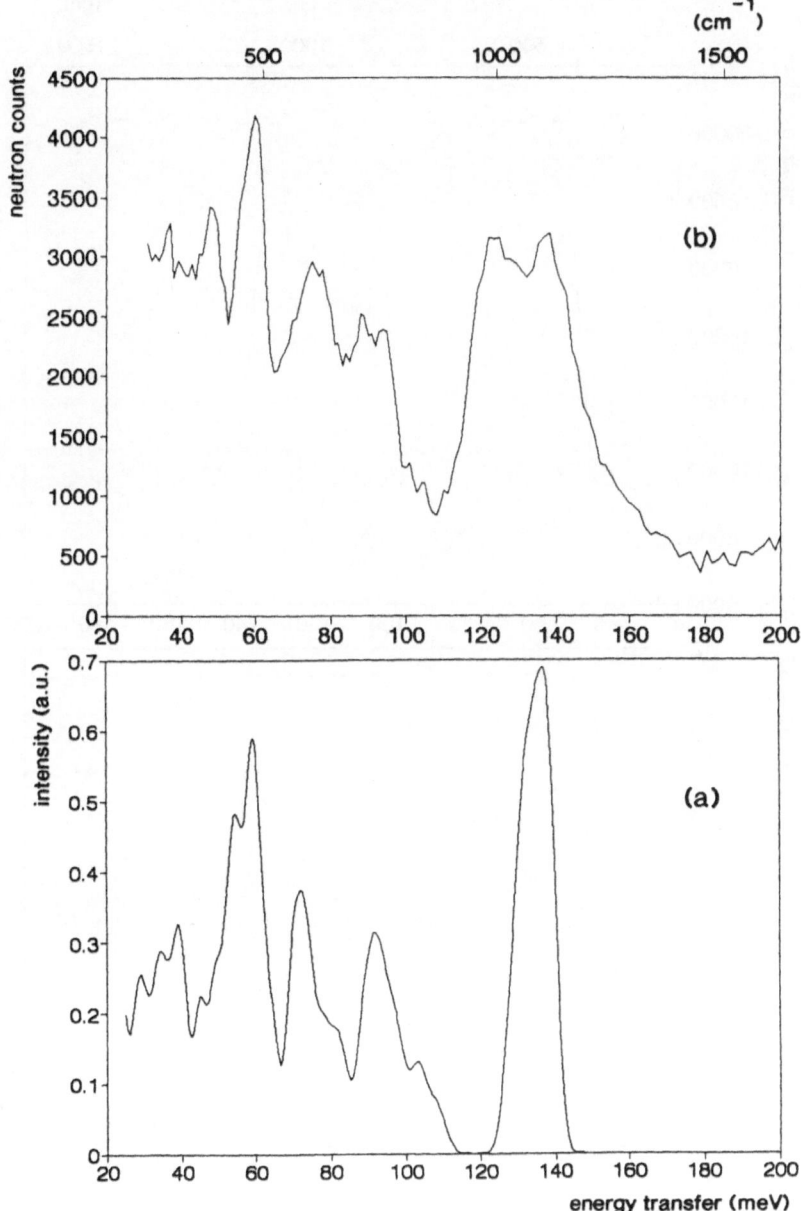

Figure 4 : INS spectra of NaY zeolite:
  (a) calculated spectrum.
  (b) experimental spectrum recorded at 20 K.

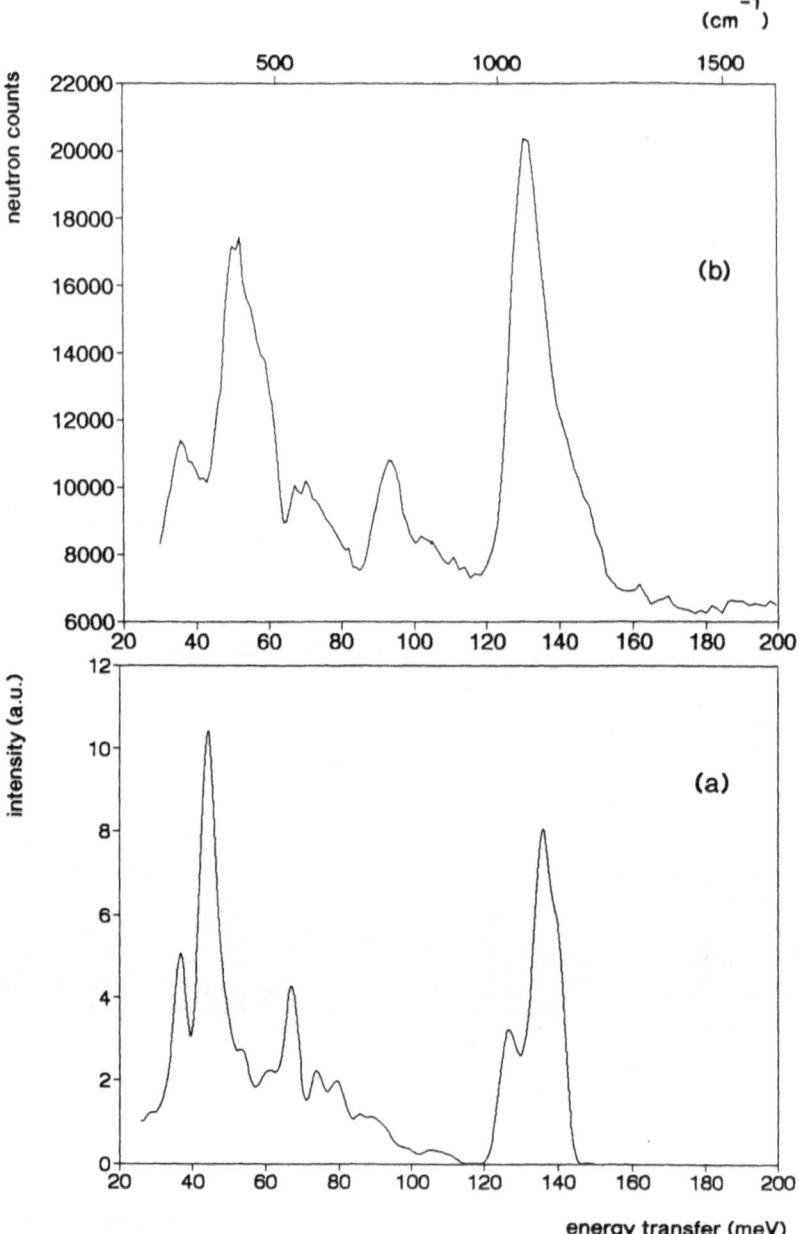

Figure 5 : INS spectra of HY zeolite:
(a) simulated spectrum.
(b) experimental spectrum recorded at 20 K.

A force field calculation, using the **GF** matrix method. was used to derive the frequencies, the displacement vectors, and the mean-square amplitudes [10]. The INS spectra were recorded with the beryllium-filter detector spectrometer INFB, at the Institut Laue-Langevin (Grenoble, France).

For NaY, the calculation was performed on a purely siliceous cluster consisting of a sodalite cage and a hexagonal prism (78 atoms). The experimental atomic coordinates were used [11] and four force constants were taken into account [10]. The calculated spectrum is shown in Figure 4(a), a reasonable agreement is observed with the experimental data of Figure 4(b), the scattering from the sodium cations being negligible. Since no electrostatic forces have been introduced, the splitting of the peak at 1100 cm^{-1} in transverse and longitudinal optical modes is not reproduced.

To calculate the INS spectrum of the protonated Y zeolite, four hydrogen atoms were added to the cluster ($2H_1$, H_2, and H_3), corresponding to the experimental occupation factors [11]. The force constants for the framework internal coordinates were not modified and four other force constants were introduced for the hydroxyl groups [10]. The calculated INS spectrum of HY in Figure 5(a) was obtained by adding up the contributions from all H, Si. and O atoms, each atom having its own Debye-Waller factor. Again, the agreement with the experimental spectrum, Figure 5(b), is satisfactory. The signal intensity is much higher in Figure 5(b) than in Figure 4(b), this is due to the large contribution from the hydrogen atoms because of the large incoherent cross section and the low mass of this element. The largest peaks near 1100 and 360 cm^{-1} in Figure 5 correspond respectively to the in-plane and out-of-plane O-H bending modes, as proposed earlier [12-13]. If these modes were decoupled from the framework. then one would expect INS intensity only in the range of the $\delta(OH)$ and $\gamma(OH)$ vibrations.

Additional peaks are found between 500 and 950 cm^{-1} in the experimental and calculated INS spectra of HY. These features have been assigned to framework vibrations and it was proposed [12] that other peaks appeared in the spectrum of HY because the hydrogen atoms were following the displacements of the framework atoms during the lattice modes. This is now supported by the simulation of the INS spectra: the intensity which is found in the spectra of HY between 500 and 950 cm^{-1} has increased compared to the spectra of NaY (note the different ordinate scales between Figures 4b and 5b). The larger intensity is therefore due to coupling of the O-H deformations with the lattice modes. The coupling is much larger for the out-of-plane than for the in-plane bending modes. The O-H stretching modes, near 3600 cm^{-1}, are pure motions and they have not been represented in Figure 5. A good simulation of the INS spectrum of a zeolitic acidic site was also obtained using the frequencies and the atomic displacements derived from ab initio calculations on small clusters [14].

INS is therefore useful to obtain detailed information on a given sample. However. due to restricted access to the neutron spectrometer. it is almost impossible to carry out a systematic study, e.g. by varying the cation composition or the zeolite structure.

4.3 Water in interaction with acidic sites in zeolites

A question which is much debated at the moment is whether the Brönsted acidity of several solids, e.g. zeolites or sulfated zirconia, is high enough to protonate water. Two possible structures have been envisaged : a hydrogen-bonded water molecule and a protonated molecule, H_3O^+, the hydroxonium ion. Most of the experimental results have been obtained by NMR or infrared spectroscopies, but the interpretation of the results is complicated. In NMR, rapid exchange can take place at room temperature between molecules in different adsorption states. In infrared, there are resonant interactions in the stretching OH region with overtones of OH deformations.

    Recent ab initio calculations performed on small clusters indicate that only the hydrogen-bonded structure is a minimum whereas the hydroxonium species is a transition structure for proton transfer [15]. However the energy difference between the two structures is small, a few kJ/mol.

    Here, the comparison between INS and theoretical calculations is fruitful. INS spectra can be simulated for the two possible water structures using the theoretical frequencies and atomic displacements as imputs [14]. The two calculated spectra are compared with the experimental data in Figure 6.

    The experimental INS spectrum, Figure 6(a), corresponds to a ratio H_2O/H^+ of 0.61 in zeolite H-ZSM-5. The spectrum was recorded with the INFB spectrometer, at the Institut Laue-Langevin. The contribution from the dehydrated zeolite has been subtracted which explains the negative peak at 1080 cm^{-1}. This band was shown in the previous section to correspond to $\delta(OH)$ deformations of the acidic groups. The negative contribution in the difference spectrum indicates interaction of water with the acid sites.

    The two calculated spectra are shown in Figures 6(b) and (c). I is worth noting that there are no adjustable parameters, only the resolution function has been introduced. It is clear from a comparison of the spectra in Figure 6 that the hydrogen-bonded water model, Figure 6(b) reproduces better the experimental profile than the hydroxonium model, Figure 6 (c). The splitting (445 cm^{-1}) and the relative intensity of the two major peaks at 455 and 900 cm^{-1} in the experimental spectrum are better reproduced by the hypothesis of hydrogen-bonded water (splitting 476 cm^{-1}) than by the assumption of a hydroxonium ion (splitting 351 cm^{-1}).

    The only deficiency in the theory is that the calculated frequencies are situated $\approx$ 80 cm^{-1} too low in energy. The assignment of the main INS bands is however simple. The band measured at 1650 cm^{-1} is assigned to the water bending. The peak at 1380 cm^{-1} corresponds to the perturbed deformation mode of the acid site, $\delta(O_zH_z)$. The perturbed $\gamma(O_zH_z)$ mode overlaps with the negative contribution

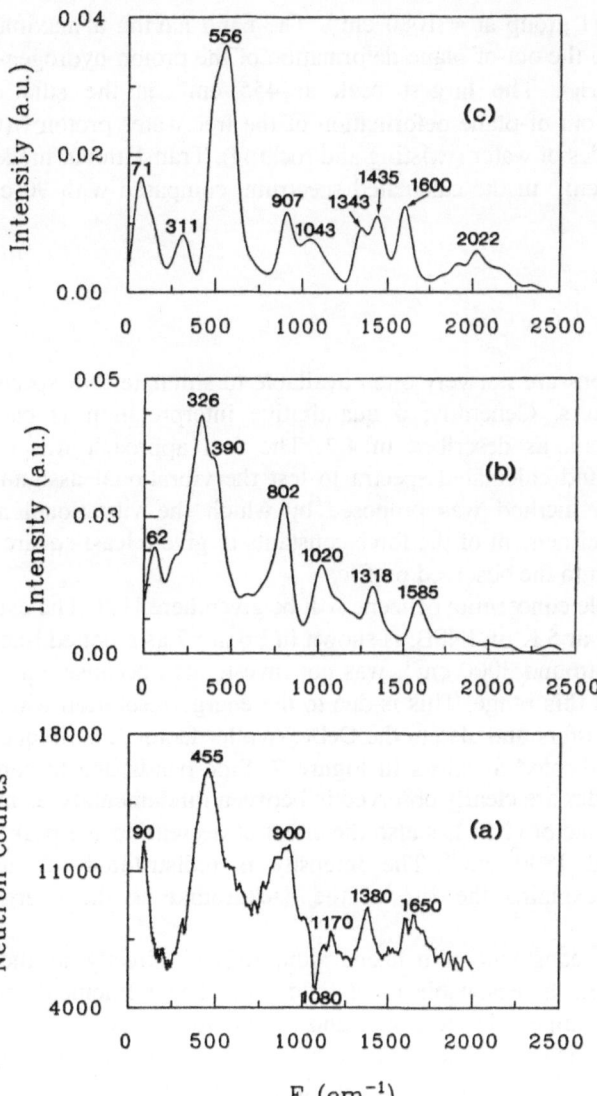

Figure 6. INS spectra of water adsorbed at low loading in H-ZSM-5
  (a) experimental spectrum
  (b) spectrum simulated for a water molecule hydrogen-bonded to a bridging hydroxyl group.
  (c) spectrum simulated for a hydroxonium ion.

of the free hydroxyl group at $\approx$ 1080 cm^{-1}. The band having a maximum at 900 cm^{-1} corresponds to the out-of plane deformation of the proton hydrogen-bonded to the zeolite : $\gamma(O_wH_b)$. The largest peak at 455 cm^{-1} is the sum of several contributions : the out-of-plane deformation of the free water proton, $\gamma(O_wH_f)$ and intermolecular modes of water (twisting and rocking). Translational modes of water give a band at 62 cm^{-1} in the calculated spectrum, compared with 90 cm^{-1} in the experiment.

4.4 Benzene

Ab initio calculations are not very often available to simulate INS spectra without adjustable parameters. Generally, a quantitative interpretation is based on an empirical force field, as described in **4.2**. The first approach was to compare visually observed and calculated spectra to test the vibrational assignment. At a later stage [16], a method was proposed by which the vibrational analysis is performed by the refinement of the force constants to give a least-squares fit of the calculated spectrum to the observed profile.

An example concerning benzene will be given here [17]. The experimental spectrum, recorded at 5 K on INFB, is shown in Figure 7 as a dotted line. The C-H stretching region, around 3000 cm^{-1}, was not investigated because only one broad band is obtained in this range. This is due to the energy resolution which worsens at large energy transfers and also to the Debye-Waller factor. The frequency of the fundamentals is indicated as sticks in Figure 7. Side bands due to combinations with the lattice modes are clearly observed in between fundamentals, as in Figure 1. The Debye-Waller factor (**2.2**) has also the effect of decreasing the peak intensities between 1200 and 1800 cm^{-1}. The intensity is redistributed in multiphonon processes, which explains the rise of the background as the energy transfer increases.

The force constants of benzene were refined directly to the observed profile, starting from a reasonable force field [18]. The refinement includes the intensities from fundamentals, overtones and combinations, the contributions from all atoms being added up. The final force field is close to the original, the intermolecular effects being small. The calculated profile is shown in Figure 7 as a continuous line. The agreement with the experimental data is reasonable. Not all the frequencies are resolved, but the method of treating overlapping modes is similar to the Rietveld method in powder diffraction. The entire INS profile is refined, instead of just adjusting the frequencies of the normal modes.

It appears that much progress has been made in the data treatment of the INS spectra, since the first quantitative interpretation of benzene [19]. In former times, the instrumental resolution was low and only the intensities from the fundamentals were calculated. The experimental spectra can be now simulated with a good accuracy. It can be noted that an even better resolution can be obtained, either on INFB but at the expense of the signal intensity [20], or on TFXA [21].

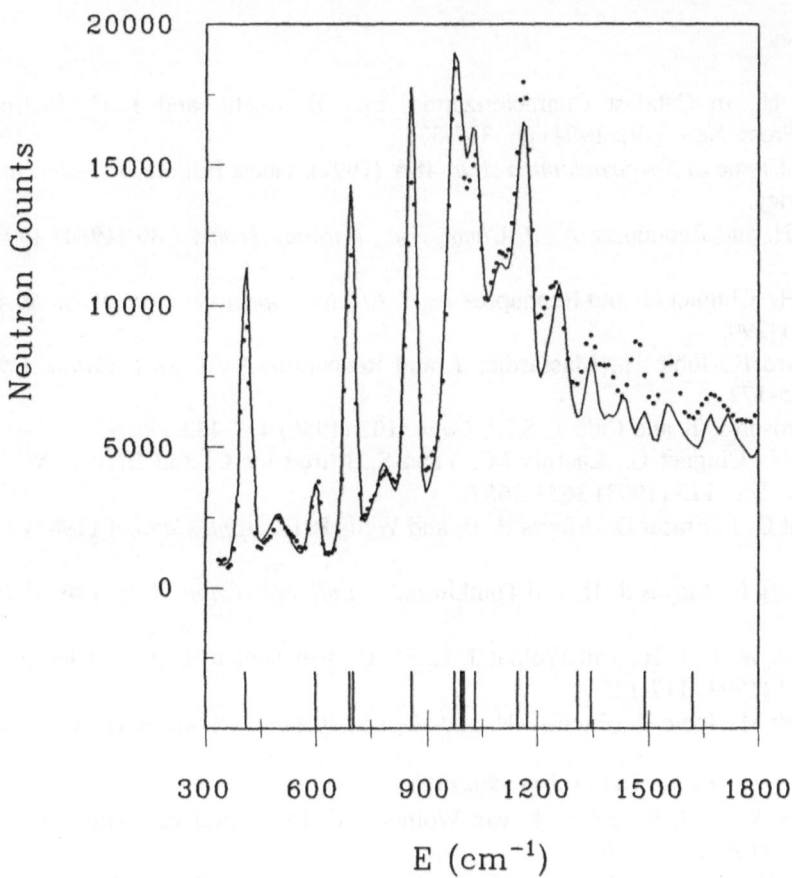

Figure 7 : Comparison of experimental and calculated INS spectra of bulk benzene.

5. CONCLUSION

In the last fifteen years, continuous progress has been made in INS, both from an experimental and theoretical point of view. The information which is extracted from the INS results cannot usually be obtained from other vibrational methods. INS is well suited to characterise different adsorbed hydrogen species or the Brönsted acidity of a catalyst. The nature of adsorbed molecules can also be identified and the adsorption geometry and strength of bonding can be determined. In the future. quantitative interpretations will be more common, to use the information contained in the intensities. However the small number of high resolution spectrometers (only four in the world) will always limit the number of applications to selected systems.

References

[1] Jobic H., in Catalyst Characterization, Eds. B. Imelik and J. C. Védrine (Plenum Press, New York 1994) pp. 347-375.

[2] Special issue of *Spectrochimica Acta* **48A** (1992), Guest Editors, J. Eckert and G. J. Kearley.

[3] Jobic H. and Renouprez A., *J. Chem. Soc., Faraday Trans. I* **80** (1984) 1991-1997.

[4] Jobic H., Clugnet G. and Renouprez A., *J. Electron Spectrosc. Rel. Phenom.* **45** (1987) 281-290.

[5] Hochard F., Jobic H., Massardier J. and Renouprez A., *J. Mol. Catal. A* **95** (1995) 165-172.

[6] Richardson J. T. and Cale T. S., *J. Catal.* **102** (1986) 419-432.

[7] Jobic H., Clugnet G., Lacroix M., Yuan S., Mirodatos C. and Breysse M., *J. Am. Chem. Soc.* **115** (1993) 3654-3657.

[8] Wright C. J., Fraser D., Moyes R. B. and Wells P. B., *Appl. Catal.* **1** (1981) 49-58.

[9] Sundberg P., Moyes R. B. and Tomkinson J., *Bull. Soc. Chim. Belg.* **100** (1991) 967-976.

[10] Jacobs W. P. J. H., van Wolput J. H. M. C., van Santen R. A. and Jobic H., *Zeolites* **14** (1994) 117-125.

[11] Czjzek M., Jobic H., Fitch A. N. and Vogt T., *J. Phys. Chem.* **96** (1992) 1535-1540.

[12] Jobic H., *J. Catal.* **131** (1991) 289-293.

[13] Jacobs W. P. J. H., Jobic H., van Wolput J. H. M. C. and van Santen R. A., *Zeolites* **12** (1992) 315-319.

[14] Jobic H., Tuel A., Krossner M. and Sauer J. *J. Phys. Chem.* **100** december 1996

[15] Krossner, M. and Sauer J., *J. Phys. Chem.* **100** (1996) 6199-6211.

[16] Kearley, G. J., *J. Chem. Soc., Faraday Trans. 2* **82** (1986) 41-48.

[17] Jobic H. and Fitch A. N., Progress in Zeolite and Microporous Materials, *Studies in Surface Science and Catalysis*, Vol. 105, H. Chon, S. K Ihm and Y. S. Uh, Eds. (Elsevier, 1997) pp. 559-566.

[18] La Lau C. and Snyder R. G., *Spectrochim. Acta* **27A** (1971) 2073-2088.

[19] Jobic H., Tomkinson J., Candy J.P., Fouilloux P. and Renouprez A. *Surface Science* **95** (1980) 496-510.

[20] Jobic H. and Lauter H. J., *J. Chem. Phys.* **88** (1988) 5450-5456.

[21] Penfold J. and Tomkinson J., Rutherford Appleton Laboratory Report RAL-86-019 (1986).

Metal Catalysis in the Conversion of Biosustainable Resources

P. Gallezot

Institut de Recherches sur la Catalyse, CNRS, 69626 Villeurbanne cedex, France

1. INTRODUCTION

Renewable or biosustainable resources are produced by living organisms, essentially by plants which convert carbon dioxide from the atmosphere into organic carbon by photosynthesis processes activated by solar light. Without including the undetermined output of marine life, biosustainable raw materials account for 2000 Gton/year of organic carbon, while only 7 Gton/year of fossil fuels (oil, coal and natural gas) are extracted from the earth. Most of these raw materials consists of cellulose and from wood, but agricultural resources also account for impressive amounts of organic carbon, e.g., 2000 Mton from cereals, 120 Mton from sucrose-containing crops and 80 Mton from triglyceride-containing crops. So far, only a very small part of these resources is used as feedstock for chemistry. However, carbohydrates derived from starch-containing cereals, or from sucrose-containing crops as well as triglycerides issued from oilseeds are increasingly processed by chemical industry (ca. 65 and 10 Mton of organic carbon from carbohydrates and triglycerides, respectively).

In spite of the present low price of fossil fuels, there are several incentives to develop industrial processes starting from renewable organic carbon. Indeed, many products derived from biosustainable resources are perceived as environmentally friendly because they are biodegradable and the intermediates or end-products processed from them are dubbed « natural » so they take a high added value particularly in the food and cosmetic industries. From a merely rational standpoint, it may be worthy to employ biomass-derived molecules because they are already functionalized in contrast with hydrocarbons which will require more steps to be converted into final products. Thus, glucose is a highly functionalized molecule which is an ideal building block for a varieties of valuable chemicals spanning from polymers to vitamin C. Furthermore, mass-produced biosustainable resources are comparatively cheap with respect to hydrocarbons, thus, the market price of technical grade glucose is similar to that of propylene.

It is not the purpose of the present review to cover all the conversion processes of renewable resources, since most of them employ non-catalytic or enzymatic processes, but rather to focus on the main processes involving metal catalysis. Indeed, metal catalysis in petrochemistry, chemistry and depollution processes is well documented, but it is less well known that metal catalysts are also used on a large scale for the conversion of agricultural products and derivatives. In this review, only the main processes using metal catalysts will be examined, namely those starting from carbohydrates and triglycerides. These aspects of metal catalysis have never been covered in a comprehensive way by books or review papers.

Figure 1 gives schematically the main processes involving metal catalysis, which start either from starch-containing crops (corn, wheat, potato) and sucrose-containing plants (sugar cane and beet) or from triglyceride-containing crops (rapeseed, sunflower, soybean). Hydrolytic or enzymatic depolymerization of starch yields glucose. Two types of metal-catalyzed glucose conversion will be examined, namely: the hydrogenation into sorbitol and the oxidation into gluconic and glucaric acids. Triglycerides, which are the triesters of glycerol and fatty acids, can be converted to edible oils and fats by partial hydrogenation of the C=C bonds on nickel catalysts. The transesterification of triglycerides with methanol gives fatty monoesters which are hydrogenated into fatty alcohols. The transesterification give glycerol as by-product which can be oxidized on metal catalysts to prepare various C_3 oxygenates.

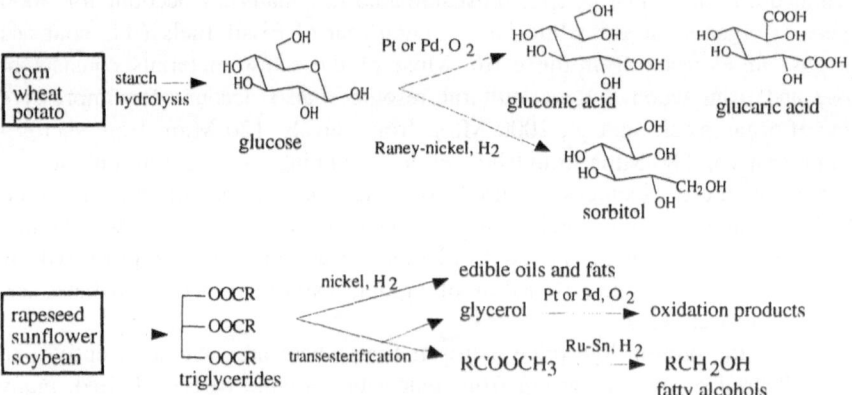

Fig. 1. Scheme of metal-catalyzed reactions

2. HYDROGENATION REACTIONS

2.1 Hydrogenation of glucose into sorbitol

D-Glucose is obtained either by hydrolysis or controlled enzymatic depolymerization of polysaccharides such as starch, cellulose or inuline, or by hydrolysis of sucrose which gives an equimolar mixture of D-glucose and D-fructose. In aqueous solutions, glucose is present under three molecular forms: two hemiacetals, α- and β-D-glucose, and a linear (acyclic) form bearing an aldehyde function. (Fig. 2). In water solutions, these molecular forms are in equilibrium with the concentration indicated. It is highly probable that catalytic hydrogenation of glucose on the metal surfaces proceeds via the terminal C=O group of the linear form. The catalytic hydrogenation of glucose accounts for an annual production of ca. 600,000 tons of sorbitol. Sorbitol (Fig. 1) is employed as a synthon for the production of various chemicals including vitamin C, as additives in foods, drugs and cosmetics as well as in the manufacture of a varieties of products, particularly paper. Its derivatives are used in protecting coatings, plasticizers, emulsifiers and detergents

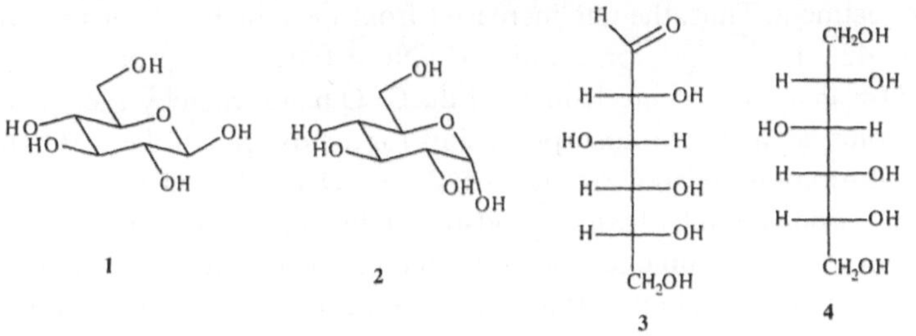

Fig. 2. Molecular forms of glucose: **1** β-D-glucose (62%), **2** α-D-glucose (38%); **3** Fischer projection of the linear form (0.002%); molecular structure of sorbitol **4**

Catalytic hydrogenation of aqueous solutions of glucose was first described early in the century by Ipatieff [1] and the first patent, issued in 1925, described the hydrogenation on kieselguhr-supported nickel catalysts [2]. Most of the early patents dealt with Raney-nickel catalysts. More recently, hydrogenation processes based on ruthenium catalysts were patented [3,4].

2.1.1 Hydrogenation on Raney-nickel and supported nickel catalysts

Raney® active metal catalysts [5], discovered by Murray Raney [6], are unsupported metal catalysts prepared by soda attack of alloys of aluminum with nickel, copper, cobalt or other transition metals. They are also known as sponge or skeletal metals and the most widely used catalysts of this type are Raney-nickel which are employed in various hydrogenation reactions, particularly glucose hydrogenation ($\approx$ 150 ton/year). Raney-nickel should be handled with care because of its pyrophoricity but has several advantages, namely: a comparatively low price, a high activity, a good resistance to poisoning (e.g., by sulfur compounds) and it can be easily separated from the reaction medium because of its high density compared to most supported catalysts. Raney-nickel catalysts employed in glucose hydrogenation contain one or more metal promoters such as molybdenum and chromium which increase the activity and stability of catalysts. Details on the preparation and characterization of typical Raney-nickel catalysts are described in appendix 1.

Hydrogenations of glucose aqueous solutions (30-40 wt%) on Raney-nickel catalysts are typically carried out batchwise, in well-stirred slurry reactors, under 3-10 MPa hydrogen pressure, at 120-150°C, with 3 to 6 wt% of catalyst with respect to glucose. Under these conditions the conversion of glucose is 100% after 2 to 3 hours and the selectivity to sorbitol is higher than 97%. The main impurity is mannitol, a stereoisomer of sorbitol which is produced by hydrogenation of fructose formed in small amounts by isomerisation of glucose.

The influence of promoter concentration on the activity and stability of Raney-nickel catalysts was recently studied [7,8]. Hydrogenation reactions were conducted on 175 ml of glucose solution (2.3 mol liter^{-1}) and 1.6 g of catalyst, at 130°C, under 4.5 MPa of H_2. Table I shows that for an optimum Cr/Ni ratio of 0.024, the initial rate is 5.5 times higher than on the unpromoted catalyst. Molybdenum was also very efficient, provided the composition of the parent alloy was homogenized by annealing treatment. Thus, the rate increased from 68 mol $h^{-1}g^{-1}$ for the unpromoted catalyst, to 431 mol $h^{-1}g^{-1}$ for a ratio Mo/Ni = 0.01. The promoting effect was interpreted by an activation mechanism of the C=O bond whereby the electropositive promoter atoms act as Lewis adsorption sites for the oxygen of the C=O group. This favors the adsorption and polarization of the C=O bonds and thus a nucleophilic attack of the carbon atom by hydrogen adsorbed dissociatively on nickel atoms. This mechanism implies an homogeneous distribution of the promoter M on the nickel surface, with an optimum concentration so that nickel surface is not entirely covered by promoter atoms. The addition of tin and iron also produced a large increase of initial reaction rates, however these catalysts deactivated rapidly, in contrast with Mo- or Cr-promoted catalysts. Thus, Fe-promoted catalyst suffered from a 80% activity loss after five recyclings because two-third of the iron atoms were leached away in solution. The deactivation of Sn-promoted catalysts was even worse because tin hydroxyde precipitate on the inner surface of Raney-nickel, blocking the access of reactant molecules to the nickel surface.

Table I. Influence of chromium concentration on the initial activities of Raney nickel

Cr/Ni (x10^3)	initial rates (mmol $h^{-1}g^{-1}$)
0	68
13	350
24	375
110	84

Raney-nickel catalysts are in powder form and thus cannot be used in continuous, fixed-bed processes. Attempts have been made to prepare extrudates or granules of Raney-nickel by wrapping them in a polymer matrix [9]. Another possibility is to replace Raney-nickel by supported-nickel catalysts. Thus, continuous processes on Ni/kieselguhr [10] or Ni/SiO$_2$-Al$_2$O$_3$ [11] catalysts have been patented. The latter catalysts contain as much as 60 wt% nickel in reduced and passivated form to avoid pyrophoricity. Glucose hydrogenation on supported nickel catalysts was recently studied in a trickle bed reactor at various temperatures, pressures and concentrations [12]. The reaction follow a Langmuir-Hinshelwood rate-law with specific activities smaller than those measured on Raney-nickel (20 to 40 mmol h^{-1}g$_{Ni}^{-1}$ compared to 70 mmol h^{-1}g$_{Ni}^{-1}$) which can be due to internal mass transfer limitation and/or to lower specific surface areas. The selectivity to sorbitol is also smaller, but the most serious limitation is the stability of the catalysts towards leaching, particularly for the support. Indeed, in addition to the fact that small amounts of gluconic acid may be formed by the Cannizaro reaction, polyols have chelating properties which account for extraction of Al and Si from supports.

In conclusion, promoted Raney-nickel catalysts are well suited for glucose hydrogenation in batch reactor as far as activity, selectivity and stability are concerned, however, discontinuous processes are not well adapted for the mass production of sorbitol which is expanding worldwide, therefore, research efforts are currently directed towards continuous processes based on ruthenium catalysts.

2.1.2 Hydrogenation on ruthenium catalysts

Ruthenium-based catalysts are well suited to hydrogenate carbonyl groups and thus are good candidates for the hydrogenation of glucose. The specificity of nickel and ruthenium compared to platinum and palladium for the hydrogenation of aldoses or ketoses was interpreted by Vasyunina et al [13] in terms of energy barriers in relation with the theory of multiplets.

Patents have been issued for the hydrogenation of glucose in continuous processes on ruthenium catalysts supported on active carbons [3] and on alumina [4]. The catalyst may deactivate because of the partial oxidation of the ruthenium surface, which can be avoided by using well outgased glucose solutions, and because of surface poisoning by deposition of iron, leached from the reactor walls, on the ruthenium surface [14]. Extensive studies on glucose hydrogenation in a trickle-bed reactor using monometallic and bimetallic ruthenium catalysts (1-3 wt% Ru) supported on active charcoal have been conducted [15]. Carbons unlike oxides were chosen as supports because there is no risk of leaching in solutions. Ruthenium was ca. 50 times more active than nickel on weight basis and the selectivity to sorbitol at total conversion was larger than 99%. Since ruthenium catalysts have better activity, selectivity and stability than nickel catalysts, the development of continuous processes based on supported ruthenium catalysts is expected in spite of the higher catalyst cost.

P. Gallezot

2.2 Hydrogenation of triglycerides into edible oils and fats

Triglycerides contained in the seeds of various crops, such as sunflower, rapeseed, and soybean, are the triesters of glycerol and unsaturated C_{12} to C_{22} carboxylic acids (fatty acids), C_{18} being the most abundant ones. Figure 3 gives a scheme of a triglyceride molecule with three different C_{18} acids: oleic (C18:1), linoleic (C18:2), and linolenic acids (C18:3) bearing one, two, and three C=C bonds, respectively. These molecules are subject to *cis-trans* isomerisation. The complete hydrogenation of any of these acids give stearic acid (C18:0).

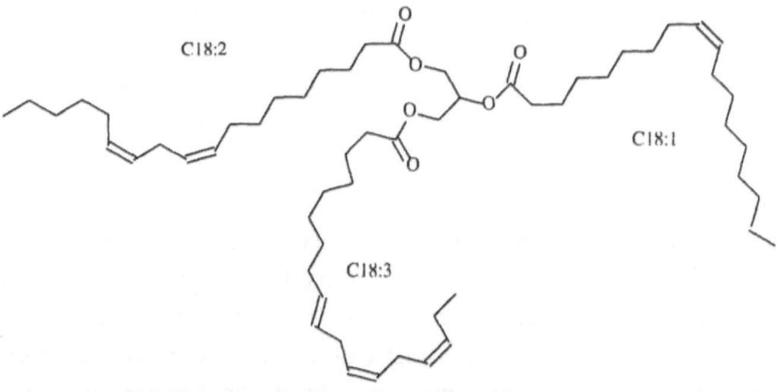

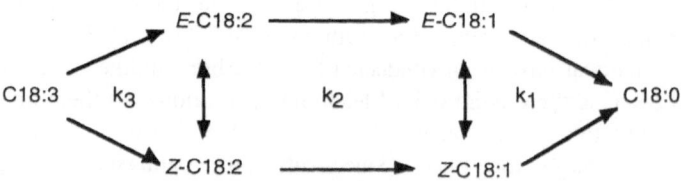

Figure 3. Scheme of a triglyceride molecule with three different fatty acids. C18:1 oleic acid; C18:2 linoleic acid; C18:3 linolenic acid.

Fatty acids are present in variable amounts in triglycerides extracted from different crops. For instance, rapeseed oil contains 50 to 60% of oleic acid, 20 to 25% of linoleic acid and 5 to 10% of linolenic acid. Partial hydrogenation of the C=C bonds is required to obtain an optimum degree of insaturation in order to control the melting point, the stability to oxidation and the organoleptic properties of edible oils and fats. The higher the number of insaturation, the lower the melting point, and the lower the stability towards oxidation.

A simplified hydrogenation scheme taking into account *cis-trans* (E-Z) isomerisation is given in figure 4. The selectivities are defined from the rate constants k_3, k_2, and k_1, as: $S_{3/2} = k_3/k_2$, $S_{2/1} = k_2/k_1$. The ideal catalyst should hydrogenate completely C18:3, produce no C18:0 and allow to control the C18:1/C18:2 ratio according to the specifications required for the final product.

Figure 4. Hydrogenation scheme of fatty esters

Hydrogenations of fatty acids are typically carried out in batch or continuous processes at 150-200°C under 2-10 MPa of H_2-pressure, on silica-supported copper, or copper-chromite, or supported nickel catalysts. Nickel catalysts supported on silica, silico-alumina or kieselguhr have a high metal loading (>20 wt%) and are coated with fully hydrogenated fats to protect them against oxidation before use. The $S_{3/2}$ selectivity is high on or copper-chromite whereas supported nickel catalysts have a high $S_{2/1}$ selectivity which is very important to achieve a good selectivity in C18.1.

These hydrogenation processes account for the production of 6 to 10 Mton/year of oils and fats which require several thousands tons of catalysts per year. Although they are continually improved, these processes have been little modified since the first quarter of the century because of the low operating costs of the current technology. However, one may expect that copper-chromite catalysts will be progressively abandoned because used catalysts containing chromium cannot be landfilled and recycling will be too costly. More details on the chemistry involved in the hydrogenation of oils can be found in the review paper of Ucciani [16].

2.3 Hydrogenation of fatty esters into fatty alcohols

Unsaturated fatty alcohols, whose production attains ca.1 Mton/year, are valuable chemicals used in the preparation of biodegradable detergents and surfactants, in cosmetics and in fine chemistry. They are prepared by hydrogenation of methyl esters obtained by transesterification of triglycerides:

$$
\begin{array}{c}
\text{—OOCR} \\
\text{—OOCR} \quad + 3\ CH_3OH \quad \xrightarrow{\ OH^-\ } \quad 3\ RCOOCH_3 + \\
\text{—OOCR}
\end{array}
\qquad
\begin{array}{c}
\text{—OH} \\
\text{—OH} \\
\text{—OH}
\end{array}
$$

$$RCOOCH_3 + 2\ H_2 \rightarrow RCH_2OH + CH_3OH$$

The carbonyl group in esters is more difficult to reduce than those of aldehydes or ketones because of the weaker polarisability of the C=O bond and the larger steric hindrance in esters. Therefore the reduction of the esters of carboxylic acids require high activation temperatures and it is a difficult challenge to hydrogenate the C=O group without hydrogenating C=C bonds. For instance, starting from methyl oleate, the formation of methyl stearate and stearyl alcohol should be avoided to maximize the yield of oleyl alcohol (Figure 5). So far, these hydrogenation reactions have been carried out with copper-chromite or zinc-chromite catalysts [17,18] at very high temperatures (250-300°C) and pressures (20-35 MPa). These processes suffer from obvious disadvantages such as severe reaction conditions and the use of environmentally unacceptable catalysts. Furthermore, they are not very selective since saturated alcohols and hydrogenolysis products are formed. These highly demanding hydrogenation processes have been improved recently using ruthenium catalysts. Desphande et al [19] claimed that a selectivity of 80% in oleyl alcohol was obtained with Ru-Sn-B/alumina catalysts at 270°C, under 4.5 MPa hydrogen. The catalysts were prepared by reduction with sodium borohydride of supports impregnated ruthenium and tin chlorides. However, the selectivity was greatly overestimated because heavy esters RCOOR', which have not been taken into account, should form by transesterification of the fatty esters (methyl oleate and methyl stearate) with the fatty alcohol formed (oleyl or stearyl alcohols). Pouilloux et al [20,21] found that Ru/Al_2O_3 or $Ru-B/Al_2O_3$ catalysts give mainly the saturated ester at 270°C under 8 MPa of H_2. In contrast, with $Ru_{2.2}Sn_{5.5}B_{0.2}/Al_2O_3$ prepared by coimpregnation and reduction with $NaBH_4$, the selectivity to oleyl alcohol attained more than 40% at total conversion, while the main side-products were heavy esters. Tin improves the selectivity, because, on the one hand, it inhibits the adsorption and hydrogenation of the C=C bond, and on the other hand, it favors the activation of the C=O group as in glucose hydrogenation on nickel catalysts (Section 2.1.1). However, an optimum concentration of tin is required because excess tin deposited on the alumina catalyzes the transesterification reactions leading to heavy esters.

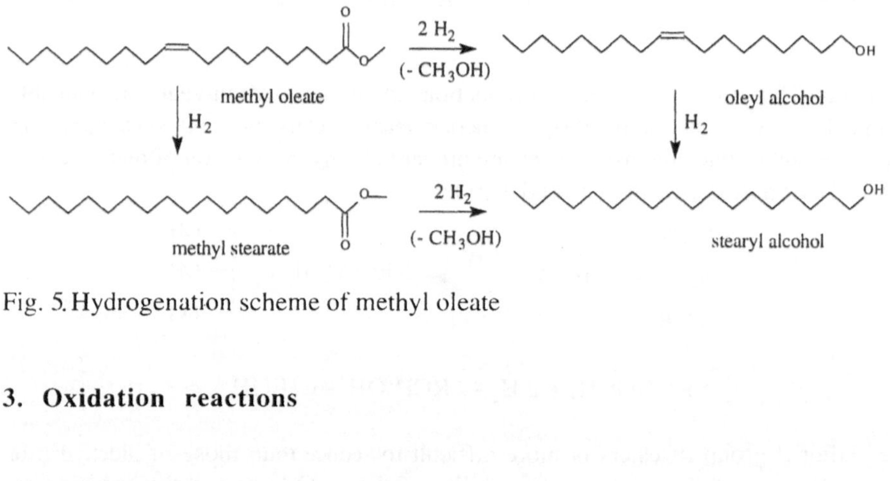

Fig. 5. Hydrogenation scheme of methyl oleate

3. Oxidation reactions

Catalytic oxidations of aqueous solutions of carbohydrates on metals were reported as early as 1861 [22]. This field of research has matured over the last fifteen years and a breakthrough was made with the discovery that the addition of metal promoters such as lead or bismuth to platinum or palladium results in profound modification of the activity and selectivity of catalysts [23-29]. Catalytic oxidation with air on supported metals can be performed slightly above room temperature with a high productivity per catalyst mass in batch or continuous mode. Generalities on liquid phase oxidation reactions with air on metal catalysts, including experimental and mechanistic aspects, are given in Appendix 2. Since it is a comparatively not well known field of metal catalysis, it is recommended to take into account these data for a better understanding of the following sections. Catalysts employed in oxidation reactions carried out in aqueous solutions are carbon-supported platinum metals. Appendix 3 gives a few examples of preparation of well defined monometallic or bimetallic catalysts supported on active charcoal or high surface area graphites.

3.1 Oxidation of glucose to gluconic acid

Gluconic acid, is a biodegradable chelating agent also employed as an intermediate in the food and pharmaceutical industry. It is produced industrially by enzymatic oxidation of glucose but an alternative route using catalytic oxidation with air on bismuth-promoted, palladium catalysts, has been patented [25]. Unpromoted palladium catalysts can oxidize glucose but the rate of the reaction decreases with conversion, and side oxidation reactions decrease the selectivity. Besson et al [30] have studied the oxidation of concentrated glucose solution (1.7 mol l^{-1}) on carbon-supported palladium catalysts with different particle size. The catalyst with particles larger than 3 nm gave complete conversion within 6 hours, whereas the conversion almost reached a plateau at two-thirds conversion on catalyst with particles smaller than 2 nm. This was attributed to a particle size dependent, oxygen poisoning of the surface, the smaller particles being the most prone to over-oxidation because of their stronger affinity for oxygen.

The beneficial effect of bismuth on the activity and selectivity was clearly demonstrated with Pd-Bi/C catalysts of homogeneous size and composition (5 wt% Pt, Bi/Pt = 0.1) prepared by deposition of bismuth on the surface of 1-2 nm palladium particles via oxido-reduction surface reactions (see appendix 3) [29]. Figure 6 shows that the rate of glucose oxidation is twenty times larger on PdBi/C (Bi/Pd=0.1) than on Pd/C.

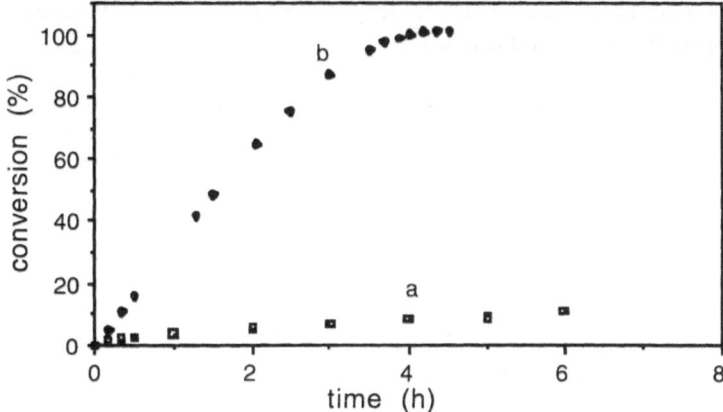

Fig.6. Oxidation with air of glucose into gluconic acid as a function of time on: (a) Pd/C; (b) Pd-Bi/C. (313 K, atmospheric pressure, pH= 9, [glucose]/[Pd] = 3150).

The selectivity to D-gluconate was also greatly improved (99.8% at 99.6% conversion) probably because the rates of side reactions were negligible with respect to oxidation into gluconate. In addition, the catalyst exhibited excellent stability since following successive recyclings there was no bismuth leaching and the activity and selectivity were almost constant, as shown by the data given in Table II.

Table II Product distribution in glucose oxidation after successive recyclings of PdBi/C [29]

Run	Conversion (%)	Yield[a] (mol %)				Selectivity (%)
		1	2	3	4	
1	99.6	99.4	<0.4	<0.4	0.2	99.8
2	99.7	98.9	<0.4	0.6	0.2	99.1
3	99.8	98.5	0.4	0.8	0.2	98.7
4	99.9	98.5	0.4	0.7	0.2	98.6
5	99.9	98.1	<0.4	0.6	0.2	99.2

Reaction conditions: time 155 min; T=313K; pH=9; glucose/Pd=787; Bi/Pd=0.1
[a]1: gluconate; 2: 2-ketogluconate; 3: 5-ketogluconate+glucarate; 4: fructose

These results were interpreted in terms of bismuth protecting palladium from over-oxidation because of its stronger affinity for oxygen, as evidenced by calorimetric measurements showing that the differential adsorption heat of oxygen was much higher on bismuth than on palladium. It was proposed that glucose oxidation proceeds according to the oxidative dehydrogenation mechanism given in figure 7 where bismuth acts as a co-catalyst.

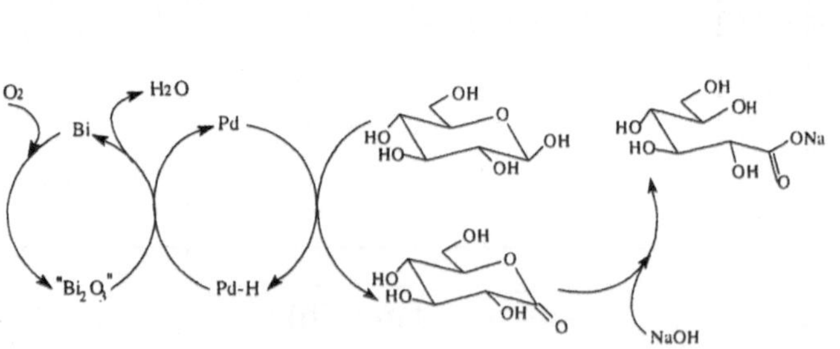

Fig. 7. Scheme of a tentative mechanism of glucose oxidation on bismuth-promoted palladium catalysts.

3.2 Oxidation of gluconic acid

Platinum catalysts can oxidize D-gluconate **1** into D-glucarate **2** or 2-keto-D-gluconate (Fig. 8), the former can be used as a biodegradable complexing agent in detergents, the latter is a valuable intermediate prepared by a fermentation process. Smits et al [23] showed that the selectivity to 2-keto-D-gluconate **3**, can be improved by lead promoter deposited on Pt/C catalysts. The selectivity to **3** was interpreted by the formation of a surface complex **4** where gluconate chelates the promoter atom via the oxygen atoms of the carboxyl group and α-hydroxyl group. The oxidation of gluconate on Pt-catalysts gives only a 55% yield of glucarate at 97.2% conversion because of the formation of more oxidized by-products such as tartarate and oxalate [31]. Platinum catalysts were prone to deactivation by over-oxidation. Reaction rates were higher after addition by surface oxido-reduction reactions of gold adatoms on the surface of platinum particles, probably because gold decreased the adsorption energy of oxygen and/or acidic side products.

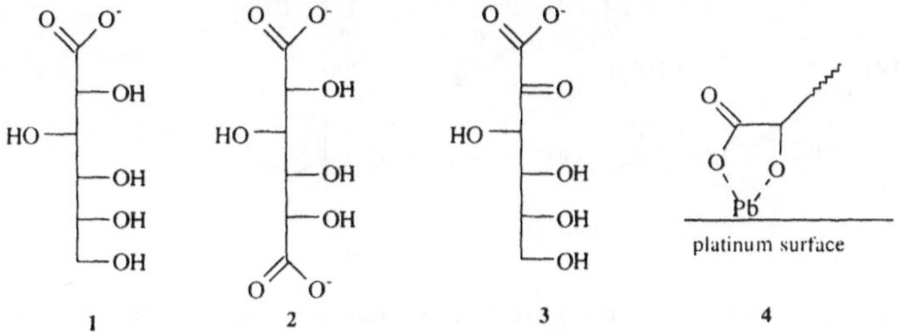

Fig. 8. Oxidation products of glucose: **1** gluconate; **2** glucarate; **3** 2-ketogluconate; **4** surface complex favoring the formation of **3**

3.3 Oxidation of 5-hydroxymethylfurfural

Fructose obtained by hydrolysis of sucrose or isomerisation of glucose is a cheap synthon available in large amounts. Its dehydration yields 5-hydroxymethylfurfural (HMF) **5** which can be oxidized into 2,5-furandicarboxylic acid **6**, a building block for polyesters or polyamides. The oxidation of HMF on Pt/alumina catalyst yielded mainly 5-formyl-2-furancarboxylic acid **7** whereas a lead-promoted Pt/C catalyst (Pb/Pt=0.25), prepared by addition of sodium hydroxide to a slurry of Pt/C in aqueous solution of lead acetate, yielded 81% of **6** (Fig. 9) [32]. The effects of reaction parameters were studied in detail and a reaction mechanism involving the complexation of lead cations by the π-electrons of the furan ring was proposed.

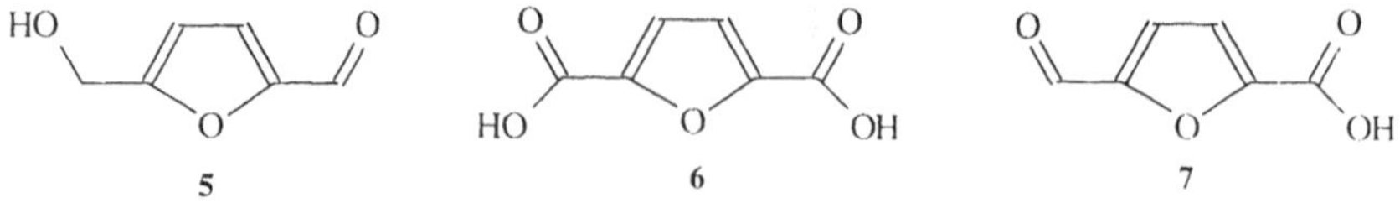

Fig.9. Oxidation products of 5-hydroxymethylfurfural

3.4 Glycerol oxidation

Glycerol is available as a by-product of the transesterification of triglycerides, therefore one can expect that its price will decrease as more biodiesel fuel will be produced in the future. It is a good candidate for use as a feedstock in the manufacture of oxygenated derivatives represented in Figure 10.

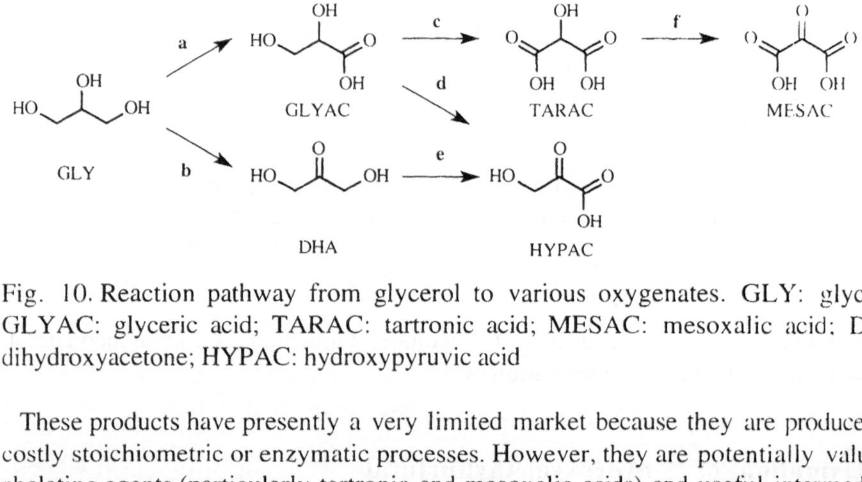

Fig. 10. Reaction pathway from glycerol to various oxygenates. GLY: glycerol, GLYAC: glyceric acid; TARAC: tartronic acid; MESAC: mesoxalic acid; DHA: dihydroxyacetone; HYPAC: hydroxypyruvic acid

These products have presently a very limited market because they are produced by costly stoichiometric or enzymatic processes. However, they are potentially valuable chelating agents (particularly tartronic and mesoxalic acids) and useful intermediates in organic synthesis. Glycerol was oxidized with air on tailored platinum or palladium catalysts where the metal surface was modified with bismuth adatoms [33-35].

In the absence of bismuth promoter or at basic pH, the primary alcohol functions are preferentially oxidized to carboxylic acids (reaction a, c, e). Thus, a 70% yield in glyceric acid was obtained from glycerol on palladium catalyst and a 83% yield in tartronic acid was obtained from glyceric acid. Secondary alcohol functions (reactions b, d, f) were oxidized selectively on bismuth-promoted platinum catalysts at acidic pH. Thus, glycerol oxidation into DHA was performed with 80% initial selectivities but the catalyst deactivated with time which may be due to the accumulation of strongly adsorbed acids on the surface. The highest DHA yield was 37% at 75% conversion. Glyceric acid oxidation proceeded rapidly to give high yields of hydroxypyruvic acid (74% at 77% conversion). Tartronic acid was oxidized to mesoxalic acid on PtBi/C at pH=1.5 with a 50% yield at 74% conversion.

3.5 Future of metal catalyzed oxidations

The few examples given above show that liquid phase oxidation with air on supported metal catalysts gives high selectivities which in certain cases such as glucose oxidation, can match or surpass those of enzymatic processes. In addition, metal-catalyzed oxidations give comparatively high productivities, e.g., up to 8 mol $h^{-1}g_{Pd}^{-1}$ for glucose oxidation on PdBi catalysts [29]. These processes offer the important advantages of high simplicity of operation (one pot reaction) and they are environmentally friendly since almost no effluents are generated.

Although mainly batch operations have been carried out so far, reactions could well be performed in continuous mode in fixed-bed catalytic reactors provided active carbon extrudates are used as supports in place of carbon powders. These reactors would provide the additional advantage of better control of the contact time of reactants and products with the catalyst which may lead to additional gains in selectivity.

For industrial applications, metal catalyst should be repeatedly recycled or used in continuous mode for a long time. This implies the absence of irreversible deactivation due to significant leaching of the metal, promoters and the supporting material. Carbon supports are particularly recommended because of their stability in acidic medium and because the formation of immobile layers of hydroxylated substrates, which cause diffusional limitation on the catalyst surface, is less probable than on oxide supports. Platinum group metals are thermodynamically stable in the zerovalent state in a wide range of conditions so that leaching could be avoided. On the other hand, bismuth atoms not directly bonded to platinum atoms are easily leached away in acidic and oxidizing conditions. With low Bi/Pt ratios and reactions carried out at controlled pH, bismuth leaching can be avoided as shown in the case of glucose oxidation [29].

4. CONCLUDING REMARKS

In this brief survey only the main processes starting from mass produced renewable resources have been mentioned. In the future, a larger number of chemicals can be derived from mass-produced biosustainable resources. Thus, the hydrogenolysis on metals catalysts of starch or cellulose yields a number of C_2-C_6 polyols which could be converted into polyamines by catalytic amination, or polyacids by catalytic oxidation. These intermediates obtained by clean catalytic processes, would be the basis for the production of polyesters and polyamides which might substitute present polymeric materials derived from hydrocarbons via cumbersome multisteps processes such as those used for the production of Nylon. Current industrial processes are mostly based on catalytic hydrogenation reactions, however catalytic oxidations on metals will probably take more importance in the near future because they are environmentally clean and present major advantages with respect to biotechnology processes, such as increased productivity and simpler installations.

Future development in this field will depend not only upon the price of hydrocarbon-derived products but on even less predictable factors such as the growing environmental concern to develop a green chemistry with renewable resources and clean catalytic processes.

APPENDIX 1

Preparation and characterization of Raney-nickel catalysts

An extensive review on preparation and physical chemistry of Raney-nickel catalysts has been achieved by Fouilloux [36]. It is not the purpose of the present section to give a complete account on the subject but merely to take as typical example, the preparation and properties of promoted Raney-nickel catalysts which were recently used in glucose hydrogenation [7,8].

Precursor alloys $Ni_{40-x}Al_{60}M_x$ (M= Mo, Cr, Fe) were obtained by melting in an induction furnace the corresponding high purity metals and the melt was cast into 1-kg ingots. All these operations were carried out in argon atmosphere to avoid metal oxidation. In some cases the ingots were annealed at 1223 K under argon to improve further the homogeneity of the solid solution of metal M in the Ni_2Al_3 lattice. The ingots were crushed into fragments and subsequently ground with a hammer-mill to obtain a fine powder which was sieved to collect only grains smaller than 40 μm. The alloy powder in 100 g batch was leached with a concentrated soda solution (500 cm^3, 6 mol liter^{-1}) added slowly at room temperature. Then the suspension was refluxed for 2 h, decanted and washed with soda (1 mol liter^{-1}). The powder was submitted to three successive refluxing treatments in 6, 4, and 2 mol liter^{-1} soda solutions and finally kept under soda (1 mol liter^{-1}) before use.

Instead of being introduced in the parent alloys, metal promoters can be deposited on the surface of unpromoted catalysts by surface redox reactions such as : $NiH + M^+ \rightarrow NiM + H^+$, or by anchoring organometallic complexes on the surface with reaction such as : $NiH + Sn(Bu)_4 \rightarrow NiSn(Bu)_3 + C_4H_{10}$. Loading of tin via organometallic complexes on Raney-nickel catalysts was described in reference [8] and similar preparation modes were reported earlier [37,38]. The composition and surface area of catalysts are given in Table III.

The high specific surface area of Raney-nickel catalysts is due to the presence of micropores formed during the soda attack of the alloy which also leads to a complete reconstruction of the structure from Ni_2Al_3 to Ni. The discrepancy between the BET area, measured by nitrogen adsorption, and the surface area, measured by thermodesorption of hydrogen, can be explained either by the presence of very small micropores which would be accessible to hydrogen but not to nitrogen, or to the presence of residual water molecules which would react with nickel during thermodesorption treatment to give hydrogen .

Table III. Characteristics of Raney-nickel catalysts

catalyst	Ni/wt%	Al/wt%	M/wt%	BET area/m^{2}g^{-1}	surface area/m^{2}g^{-1}
RNi	90.6	9.1		77	106
RNiCr[a]	83.4	13.5	1.8	100	146
RNiMo[a]	84.2	11.6	1.24	83	113
RNiSn[b]	80.5	8.3	5.0	70	41

[a] prepared from Ni-Al-M alloys; [b] prepared by surface reaction on RNi with Sn(Bu)$_4$

Molybdenum- and chromium-promoted catalysts have higher surface areas, particularly in the case of chromium. The pore size distribution, measured from nitrogen adsorption isotherms, indicates a much higher fraction of pore smaller than 2 nm. X-ray line broadening analysis indicates that the average particle sizes are 3.6 and 4.4 nm in RNiCr and RNi, respectively.

APPENDIX 2

Liquid phase oxidations with air on metal catalysts

A 2.1 Reaction conditions and catalysts
Oxidation with air of aqueous solutions of alcohols, polyols, and aldehydes, are usually carried out in well-stirred batch reactors in the presence of a suspension of the catalyst in powder form. Reactions are run at atmospheric pressure under continuous stirring with air bubbling through the suspension maintained at constant temperature in the range 20 to 80 °C. Activities and selectivities are strongly dependent on the pH; a constant pH can be regulated by the addition of dilute alkali solutions via a pump under the control of a pH regulator. It is also useful to check if the reaction kinetics are under mass transfer control by monitoring the oxygen pressure in the liquid phase with an oxygen sensor. A platinum electrode associated with a reference electrode can be used to measure the electrochemical potential of the catalyst providing in situ information on the oxidation state of the active metal surface, which may be used to optimize the oxygen supply to the reaction medium[39].

The initial conditions for starting the reaction (e.g., the contact time of the catalyst with the solution under nitrogen atmosphere) can greatly affect the activity because different pretreatments modify the initial oxygen coverage of the surface (vide infra). In most cases, catalysts consist of palladium or platinum particles (size: 1 to 10 nm) supported on active carbons. These supports present the advantage of high stability under all reaction conditions, particularly at low pH and in the presence of complexing molecules. A large variety of metal catalysts, including bismuth-promoted ones [40], are available commercially, but tailor made catalysts may also be prepared to meet specific needs, for example, optimizing the metal dispersion or distribution in catalyst grains (eggshell or uniform distribution). Appendix 2 gives some examples of preparation of well defined monometallic or bimetallic platinum catalysts supported on active charcoal and high surface area graphites.

A 2.2 Mechanism of carbohydrate oxidation

It was suggested, at a very early stage, that oxidations of carbohydrates on metal surfaces proceed via a dehydrogenation mechanism [41]. This is supported by the fact that isotopically-labelled oxygen used as the oxidizing agent is not incorporated into reaction products [42] and by measurements showing that the electrochemical potential of the platinum catalysts in alcohol solutions is almost similar to that of the hydrogen electrode, which means that the Pt-surface is covered by adsorbed hydrogen [43]. Similar measurements during oxidation reactions also showed that the surface is partially covered with hydrogen [44]. Also, glucose is dehydrogenated on platinum catalysts at room temperature and basic pH, yielding gluconate and molecular hydrogen which evolves from the reaction medium [45]. If there is a general agreement on the dehydrogenation mechanism, the precise reaction pathway is still under debate. Indeed, there are still too many uncertainties as to the nature and coverage of the adsorbed species on the metal surface (e.g., H or $H^{(-)}$, O or $OH^{(-)}$, RCH_2OH or RCH_2O^- or RCH_2O.) which depend upon many factors such as pH, oxygen availability, carbohydrate concentration and nature of the metal.

A 2.3 Catalyst deactivation and role of promoters

Although the oxidation of alcohols and polyols on metal catalysts proceed at low temperatures, a deactivation due to metal sintering could occur by a mechanism of transport of metal atoms from the small to the large particles by reactants molecules which have chelating properties. Metal leaching, particularly of electropositive metal promoters should be of serious concern, especially when large amounts of these promoters are used. However, the main cause of deactivation is the oxygen poisoning of the metal surface. The current oxygen coverage of the surface will depend upon the relative affinity of the metal for oxygen and for the organic substrate; dehydrogenation of the latter on the metal surface yields dissociated hydrogen which scavenges stoichiometrically the chemisorbed oxygen. This dynamical balance of competitive adsorption controls the reaction rate both initially, and as reaction proceeds. The equilibrium tends to shift towards an overwhelming oxygen coverage as the concentration of the substrate in solution decreases, which poisons the reaction. Note that the presence of strongly adsorbed reaction products or by-products blocking part of the surface may produce a genuine deactivation which ultimately favors surface over-oxidation resulting in a much greater deactivation. Various factors affect the initial rate and subsequent deactivation, some of them are given below.

Factors depending on the catalyst : (a) Metals with a higher redox potential will be less prone to oxidation [46]. In that respect, among Pt-group metals, platinum catalysts will be the less easily poisoned by over-oxidation, followed by palladium. (b) Small metal particles (e.g., < 2 nm) deactivate more readily than larger ones, as shown in the study of D-glucose oxidation on Pd/C catalysts (see section 3.2) where the particle size dependent deactivation was attributed to the stronger affinity of oxygen for small particles [30]. (c) Mass transfer limitation in catalyst pores may help prevent deactivation [47]; thus, as the concentration of oxygen decreases continuously from the edge to the core of the catalyst grain, there is always a zone at a certain depth, where the concentration of oxygen is low enough to avoid metal surface over-oxidation.

Factors depending on the species in solution. (a) The higher the affinity of the carbohydrate for the metal, the lower the oxygen coverage; thus, platinum-group metals were not poisoned in the oxidation of 5-hydroxymethylfurfural because of strong bonding of the substrate via the π-electrons of the furan ring [48]. (b) The higher the reduction potential of the substrate, the lower will be the deactivation. (c) Initial activity as well as deactivation depend markedly upon the pH: lower rates and high deactivation are observed for reactions at acidic pH because undissociated carboxylic acids are very strongly adsorbed on metals. They can poison the reaction, directly by blocking the surface, and indirectly by favoring over-oxidation processes.

Factors depending on the reaction conditions. The higher the oxygen pressure in the liquid phase, the greater the risk of over-oxidation. If the initial rate of oxidation is fast enough, oxygen dissolved in the aqueous solutions will be totally consumed (the reaction rate is then limited by gas-liquid oxygen mass transfer) and the risk of over-oxidation is weak. For sluggish reaction, over-oxidation can be prevented by working at low and constant concentration of oxygen by dilution with nitrogen or by taking advantage of mass transfer limitation in pores (vide supra). Monitoring the electrochemical potential of the catalyst to control the optimum oxygen supply is a good way to cope with this problem [49].

APPENDIX 3

Preparation of carbon-supported catalysts

Activated carbons are widely used as catalyst supports in liquid phase processes involving intermediates and fine chemicals particularly those produced from biosustainable resources. They are well suited because of their high specific surface area and organophillicity but the main interest is their strong resistance in almost any reaction media, particularly in the presence of acidic and/or complexing agents. They are also well suited for easy recovery of metals from spent catalysts since a simple combustion release all mineral components. The most usual carbon supports are active charcoals (up to 2000 m^2g^{-1}) produced by physical or chemical activation of wood or peat but materials derived from fruit kernels or coconut shells are also used whenever a particular porosity is required. Most of the active charcoals are available in powder form with optimum particle sizes, i.e., without too small grains which would not settle in liquids, or too large grains which would decrease the effectiveness of the inner active sites. Some active carbons are available in extrudate form (0.5-3 mm) for application in fixed-bed, continuous reactors. Synthetic carbon blacks of high surface (up to 300 m^2g^{-1}) without microporosity would be interesting supports for processes requiring the absence of micropores but particles are too fine for practical use since they remain in suspension. Synthetic high surface graphites (up to 300 m^2g^{-1}) have also a low microporosity and the graphite structure exerts an electronic support effect on small metal aggregates which can modify their catalytic properties (vide infra).

Metals can be loaded on carbon supports by conventional impregnation or coimpregnation methods followed by H_2-reduction. However, impregnation techniques often result in poorly dispersed samples specially when aqueous impregnation solutions are employed since they do not penetrate easily in the micropores because of the support hydrophobicity. Ion-exchange is a very reproducible method to prepare small metal aggregates uniformly distributed on active carbons and graphites. A typical ion-exchange procedure to load platinum group metals is described below.

The carbon supports are oxidized to create exchangeable carboxylic acid groups by stirring 20 g batches in 500 ml sodium hypochlorite solutions (15 % active chlorine) for 24 h at room temperature. After filtration, the carbons are washed with 500 ml of HCl (1 mol liter^{-1}), then with water until neutrality of the wash-waters, and dried overnight at 373 K under reduced pressure. The functionalized supports are ion exchanged with aqueous solutions of the ammino cations of platinum metals such as $Pt(NH_3)_4^{2+}$, $Pd(NH_3)_4^{2+}$, $[Ir(NH_3)_5Cl]^{2+}$, $[Rh(NH_3)_5Cl]^{2+}$, and $Ru(NH_3)_6^{3+}$. The exchange is performed by stirring 20 g of the support for 18h under nitrogen atmosphere in ammoniacal solutions containing the required amounts ammino cations. The suspension is filtered, washed with water, and dried overnight at 373 K under flowing nitrogen atmosphere. Reduction of metal cations is carried out by heating at 1 K min^{-1} from 298 to 573 K under a flow of hydrogen and maintaining this temperature for 2h. The reduced catalysts are cooled to 300 K under argon and finally brought into contact with air diluted with argon to avoid deep metal oxidation. During these treatments the protons of the carboxylic groups are exchanged by the metal cations. This process occur homogeneously over all the available sites and the exchange level is limited only by the number of these sites. Upon reduction, the metal atoms remain anchored to the functional groups which results in uniformly distributed very small aggregates (typically 1nm for ruthenium and between 1 and 2 nm for other Pt-group metals. In the case of active carbons, the aggregates are distributed throughout the micropores, whereas on graphite the particles decorate the graphite steps since the functional groups created by oxidation treatment are located at the extremities of graphite basal planes [50-54]. It was shown that there is an electron transfer from graphite to the aggregates producing an expansion of the metal lattice and of a modification of the catalyst selectivity in the hydrogenation of unsaturated carbonyl compounds.

Bimetallic aggregates can be prepared by co-exchange; thus, Pt-Ru aggregates of uniform composition were prepared by coexchange and reduction [55]. Bimetallic particles can also be prepared by a surface oxido-reduction processes. Thus, iron or ruthenium atoms were deposited on carbon-supported catalyst, by adding a metal salt on a suspension of Pt/C saturated with hydrogen [56]. Similarly, palladium-bismuth catalysts were prepared by introducing the required amounts of $BiONO_3$ in a suspension of Pd/C catalysts in glucose solution [29]. Under these conditions bismuth adatoms were deposited on the surface of the metallic particles by a redox surface reaction:

$$(Pd-H)_{surface} + BiO^+ \rightarrow (Pd-Bi)_{surface} + H_3O^+$$

The homogeneity of the composition of bimetallic particles was verified by STEM-EDX analysis at 1.5 nm spatial resolution. These techniques which can be also used to obtain an homogeneous growth of metal particles as described by Menezo et al [57]. Thus, carbon black-supported platinum catalysts used for fuel-cell electrodes were obtained with controlled particles sizes [58]. Electrodeless deposition techniques are very efficient to obtain homogeneous composition or growth on carbons supports because the adatoms, particularly electropositive ones, are deposited very selectively on the metal surface rather than on the support.

References

[1] Ipatieff V.J., *Ber. Deutsch. Chem. Ges.*, **5** (1912) 3218.
[2] German Patent 544 666 (for I.G. Farben).
[3] US Patent 2 868 847 (1959, for Engelhard Industries).
[4] US Patent 4 413 152 (1983), 4 471 144 (1984), 4 503 274 (1985), 4 510 339 (1985) (for UOP).
[5] Registred name for W.C. Grace & Co.
[6] US Patent 1 563 787 (1925); 1 628 191 (1927); 1 915 (1933).
[7] Cerino P.J., Flèche G., Gallezot P., and Salomé J.P., « Heterogeneous Catalysis and Fine Chemicals », M. Guisnet, J. Barrault, C. Bouchoule, D. Duprez, G. Perot, R. Maurel, and C. Montassier Eds., (Elsevier, Amsterdam, 1991), p. 231.
[8] Gallezot P., Cerino P.J., Blanc B., Flèche G., and Fuertes P., *J. Catal.*, **146** (1994) 93.
[9] US Patent 4 826 799 (1989, for W.R. Grace).
[10] US Patent 2 759 024 (1956, for Atlas Powder Co.).
[11] US Patent 4 322 569 (1982, for Hydrocarbon Research).
[12] Déchamp N., Gamez A., Perrard, A., and Gallezot P., *Catalysis Today*, **24** (1995) 29.
[13] Vasyunina N.A., Barysheva B.S., Balandin A.A., *Izv. Akad. Nauk SSSR, Ser. Khim.*, **4** (1969) 848.
[14] Arena B.J., *Appl. Catal.*, **87** (1992) 219.
[15] Déchamp N., Perrard A., and Gallezot P., unpublished results.

[16] Ucciani E., « Heterogeneous Catalysis and Fine Chemicals », M. Guisnet, J. Barrault, C. Bouchoule, D. Duprez, C. Montassier, and G. Perot Eds., (Elsevier, Amsterdam, 1988), p. 33.

[17] Kreutzer U.R., *J. Am. Oil. Chem.*, **61** (1984) 343.

[18] Boerma H., « Preparation of Catalysts », B. Delmon, P.A. Jacob and G. Poncelet Eds., (Elsevier, Amsterdam, 1976), p. 105.

[19] Desphande V.M., Ramnarayan K., and Narasimhan, *J. Catal.*, **121** (1990) 174.

[20] Pouilloux Y., Piccirilli A. , and Barrault J., *J. Mol. Cat. A: Chemical*, **108** (1996) 161.

[21] Piccirilli A., Poilloux Y., Pronier S., and Barrault J., *Bull. Soc. Chim. Fr.*, **132** (1995) 1109.

[22] von Gorup-Besanz E., *Ann.*, **118** (1861) 257.

[23] Smits P.C.C., Kuster B.F.M., van der Wiele K., and van der Baan H.S., *Appl. Catal.*, **33** (1987) 83.

[24] Abbadi A., and van Bekkum H., *Appl. Catal.A*, **124** (1995) 409.

[25] Fuertes P., and Flèche G., Roquette Frères, EP 233816 (1987).

[26] Brönnimann C., Bodnar Z., Hug P., Mallat T., and Baiker A., *J. Catal.*, **150** (1994) 199.

[27] Verdeguer, P. Merat N., and Gaset A., *J. Mol. Cat.*, **85** (1993) 327.

[28] Gallezot P., Besson M., and Fache F., « Catalysis of Organic Reactions », M.S. Scaros and M.L. Prunier Eds., (Marcel Dekker , New York, 1995), p. 331.

[29] Besson M., Lahmer F., Gallezot P., Fuertes P., and Flèche G., *J. Catal.*, **152** (1995) 116.

[30] Besson M., Gallezot P., Lahmer F., Flèche G., and Fuertes P., « Catalysis of Organic Reactions », J R Kosak, T A Johnson Eds., (Marcel Dekker, New York,1993), p.169-180.

[31] Besson M., Flèche G., Fuertes P., Gallezot P., and Lahmer F., Recl. Trav. Chim. Pays- Bas, **115** (1996) 217.

[32] Verdeguer P., Merat, N. and Gaset A., *J. Mol. Cat.*, **85** (1993) 327.

[33] Garcia R., Besson M., and Gallezot P., Appl. Catal. A, **127** (1995) 165

[34] Fordham P., Besson M., and Gallezot P., Appl. Catal. A, **133** (1995) L179

[35] Fordham P., Garcia R., and Besson M., andGallezot P., « Proceedings of the 11th International Congress on Catalysis », J.W. Hightower and W.N. Delgass, E. Iglesias, and A.T. Bell Eds., (Elsevier, Amsterdam, 1996), p. 161.

[36] Fouilloux P. *Appl. Catal.* **8** (1983) 1.

[37] Burnonville J.P., Candy J.P., and Mabilon G., US Patent 4 628 130 (1986).

[38] Margitfalvi J. L., Göbölös S., Tals E., and Hegedüs M., « Preparation of Catalysts IV », G. Poncelet, P.A. Jacobs, P. Grange, and B. Delmon, Eds., (Elsevier, Amsterdam, 1991), p. 669.

[39]. Mallat T., and Baiker A., *Catalysis Today*, **19** (1994) 247.

[40] Despeyroux B.M., Deller K., Peldszus E., « New Developments in Selective Oxidation », G. Centi and F. Trifiro Eds (Elsevier, Amsterdam, 1990), p.159.

[41] Heyns H., Paulsen H., Ruediger G., and Weyer J., *Fortschr. Chem. Forsch.*, **11** (1969) 285.

[42] Rottenberg M., and Baertschi, *Helv. Chim. Acta*,, **227** (1956) 1073.

[43] Müller , and Schwabe K., *Z. Electrochim. Acta*, **33** (1928) 170.

[44] Mallat T., Bodnar Z., and Baiker A., « Hetrogeneous Catalysis and Fine Chemicals III » M. Guisnet et al Eds, (Elsevier, Amsterdam, 1993), p 377.

[45] de Wit G., de Vlieger J.J., Kock-van Dalen A.C., Heus R., Laroy R., van Hengstum A.J., Kieboom A.P.G., and van Bekkum H., *Carbohydr. Res.*, **91** (1981) 125.

[46] Gallezot P., de Mésanstourne R., Christidis Y., Mattioda G., and Schouteeten A., *J. Catal.*, **133** (1992) 479.

[47] van Dam, H.E., Duijverman P., Kieboom A.P.G., and van Bekkum H., *Appl. Catal.*, **33** (1987) 373.

[48] Vinke P., van Dam H.E., and van Bekkum H., « New Developments in Selective Oxidation », G. Centi, and F. Trifiro, Eds., (Elsevier, Amsterdam, 1990), p.147.

[49] Mallat T., and Baiker A., *Catalysis Today*, **24** (1995) 143.

[50] Richard D., and Gallezot P., « Preparation of Catalysts IV », B. Delmon, P. Grange, P.A. Jacobs and G. Poncelet, Eds, (Elsevier, Amsterdam, 1985), p. 71

[51] Giroir-Fendler A., Richard, D., and Gallezot, P., « Heterogeneous Catalysis and Fine Chemicals », M. Guisnet, J. Barrault, C. Bouchoule, D. Duprez, C. Montassier and G. Perot, Eds., (Elsevier, Amsterdam, 1988), p. 171.

[52] Richard, D., Gallezot, P., Neibecker,D., and Tkatchenko, I., *Catal. Today* 3, 53 (1989)

[53] Richard, D., Bergeret, G., Leclercq, C., and Gallezot, P., *J. Microsc. Spectrosc. Electron.* **14** 377 (1989).

[54] Gallezot, P., Richard, D., and Bergeret, G., « Novel Materials in Heterogeneous Catalysis », R.T.K. Baker and L.L. Murrel Eds, (Am. Chem. Soc., Washington DC, 1990), p. 150.

[55] Giroir Fendler A., Richard D., and Gallezot P., *Faraday Disc.*, **92** (1991) 69.

[56] Richard D., Ockelford J., Giroir Fendler A., and Gallezot P., *Catal. Lett., 3* (1989) 53.

[57] Menezo, J.C., Denanot, M.F., Peyrovi, S., and Barbier, J., *Appl. Catal.* **15**, 353 (1985).

[58] Gamez, A., Richard, D., Gallezot, P., Gloaguen, F., Faure, R., and Durand, R., *Electrochimica Acta* **41**, 307 (1996)

Imprimé en France. — JOUVE, 18, rue Saint-Denis, 75001 PARIS
N° 250077T. Dépôt légal : Octobre 1997